Australia

A guide and resource fo:

C000220238

William James Davis, Ph.D.

Copyright © 2009 by Wm. James Davis
ISBN 978-0-9822654-1-3
0-9822654-1-7
http://www.TerraNat.com

Copyright © 2009 by William James Davis. All rights reserved. No part of this book may be used or reproduced in any manner whatsoever without written permission except in the case of brief quotations embedded in critical articles and reviews. To request permission, the publisher can be reached by visiting www.terranat.com

Cover drawings: Australian Magpies (front) and nesting Magpie-Larks and Willy Wagtails (top back) by Marilyn Rose; White-winged Choughs (center back) by Aremy McCann. All photos on covers by Gail R. Hill with the exception of the Mistletoebird by David Stowe. Photos on covers from left to right: Galah, Brush-turkey, Figbird, Mistletoebird, Eclectus Parrot, Superb Fairy-wren, Southern Cassowary

This book is dedicated to Gail R. Hill, a fellow naturalist and dear friend who has provided many of the excellent photos and has freely shared her observations and ideas.

Also by William James Davis

Terra Explorer Vol. 1: A resource for naturalists and nature video journalists:

Table of Contents

Introduction

Behavioral profiles

Adaptive attributes

Plumage colors and patterns

Vocalizations and sounds

Breeding and nesting

Response to predators

Introduction

"Oh sure, crèching is in the book.
But that isn't in here."

Australian landscapes vary from tropical rainforest to some of the driest deserts in the world. This island continent is surrounded by thousands of miles of coastline which abut deserts, rainforests, and urban centers. High habitat diversity, long-term isolation, and unpredictable rainfall patterns have set the stage for the evolution of unique flora and fauna. As a consequence, many endemic species exhibit behaviors found nowhere else in the world.

With regard to birds, many Australian species are as astounding to watch as they are beautiful. The dawn chorus of Laughing Kookaburras, the courtship displays of Satin Bowerbirds, and the mound building of Brush-turkeys are three of many examples. Field guides help identify plants and animals; this book will help you recognize and appreciate the unique behavioral ecology of Australian birds.

In the first section, you will find 29 behavioral profiles of Australian birds. In many profiles, a species' most noteworthy behaviors are described and illustrated in black and white line drawings. In some profiles, distinctive behaviors are discussed in detail, such as tail-wagging by Willy Wagtails, kidnapping by White-winged Choughs, and petal-carrying by Superb Fairy-wrens.

In the second section of the book, titled "Adaptive attributes," the emphasis is on introducing basic facts that relate to the social and ecological pressures directing the evolution of behavior. Such information can be extremely helpful in understanding why Australian birds behave the way they do.

Terra Explorer

Watching how birds behave provides a good reason to enjoy the outdoors but also offers an opportunity to learn how different species cope with unique challenges in their physical and social environments. By just remaining keenly alert, you will undoubtedly make interesting and delightful discoveries, even some that are new to science. This book was written to help you interpret the behaviors that you witness while watching birds. But truthfully, a single guide can only offer a limited amount of information. Do not despair, however, because it is the thrill of encountering the unknown that adds a sense of adventure to field excursions. To maximize your pleasure and productivity, I recommend recording your observations on videotape. Doing so will give you the opportunity to relive your experiences and, in the process, inspire new insights. Of equal importance, when events are recorded on tape you

1

can share your discoveries with other naturalists at home and over the internet.

The material in this book presents ideas concerning what to video-tape and contextual information that will help you interpret your observations. If exploring with a camcorder appeals to you, I suggest that you consider uploading your best stories and videos to the **Terra Explorer Project** located at www.terranat.com. In fact, a few of the behaviors mentioned in this book have been videotaped and described online.

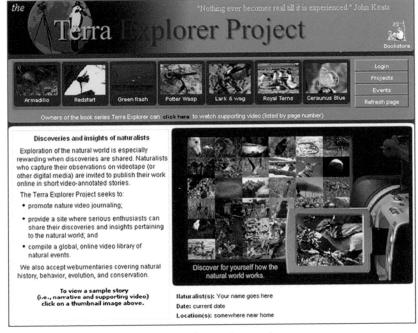

Homepage of www.terranat.com

"I find studying the behaviour of animals in their natural surroundings a fascinating hobby. It allows one to live out of doors and in beautiful scenery; it gives free scope to one's urge to observe and to reflect; and it leads to discoveries. Even the most trivial discovery gives intense delight."

Niko Tinbergen, Nobel laureate

2

Behavioral profiles

"You know, you're right. It is the Macarena!"

Emu

(height up to 2 m)

WJD

Dromaius novaehollandiae

Breeding

Emus have a long reproductive cycle—up to 2 years from pairing to independence of the offspring. They breed at most latitudes across Australia, except in northern Queensland. Sexual dimorphism is most evident during the breeding season when females develop a large pouch and fluffy feathers at the base of the neck. Also, the skin at the top of the neck of both males and females turns a turquoise blue (but not in all adults).

Female Emus preferentially court a specific mate, whereas the male accepts most females in his vicinity. Females are also more aggressive and defensive than males, suggesting that there is competition between females for access to males. DNA fingerprinting indicates lots of extra-pair copulations.

LS

Standing tall: During a prolonged confrontation, Emus will stretch their body upward in an attempt to become as tall as possible. If neither bird backs down, one opponent may charge its rival. Notice the slightly inflated throat pouch of the bird on the right.

During courtship, females emit a resonant boom or drumming sound, whereas males produce a short, coarse grunting sound. While displaying, both sexes fluff out the feathers on the front of the neck which is curved into an S-shaped posture. Also, a female is likely to perform a *high step* walk that drifts to one side as she progresses.

Some females produce a very strong odor to attract and activate the sexual interests of males. When a male prefers a specific female, he follows her and attempts to place his head over her back or nape. Females exhibit sexual interest by lowering the head, raising the tail, and crouching in the mating position. A responsive male crawls toward her and introduces his erect phallus into her cloaca. Incidentally, very few birds have a phallus.

Broody behavior

An Emu's nest is little more than a shallow scrape on a bare patch of ground. While a female lays an egg, the nest's owner is often standing nearby all the while tossing small twigs and leaves over his back. As many as 15 eggs can be laid in a nest, but eventually an upper limit is reached when the attending male pushes away any females that approach the nest. A male may lose up to 25 percent of his body weight while sitting on the eggs during the continuous 56-day incubation period. Upon hatching, the male cares for the chicks for another 18 months.

The striped pattern of Emu chicks makes the young birds difficult to see when they hide under shrubs or feed among mallee vegetation. In the wild, several broods may join to form a crèche that is attended by an adult male.

Aggression

Grunting and hissing accompany most aggressive interactions. During prolonged exchanges, Emus stretch their body upward to become as tall as possible. If neither bird backs down, one opponent may charge its rival and a chase ensues. During a fight, Emus jump up and kick forward and down with their toes spread open.

Females are aggressive toward other females that attempt to court their partner. Some males are defensive during incubation, but males are mostly aggressive when they are with their chicks. On most occasions, aggressive Emus will back off if a human raises an arm above the Emu's head—using a stick as an extension of the arm is useful when an Emu is very tall!

Strutting: When sexually active, Emus, particularly females, strut about with their throat pouch inflated. The pouch is used as a resonating chamber to amplify the Emu's territorial booming.

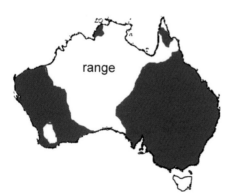

Habitat

Emus occur in a wide range of habitats, from desert to woodlands, but rarely in rainforests. Their preferred habitat is sclerophyll forests.

Australian Brush-turkey

Breeding

While the eggs of most birds are laid in a snug nest and then warmed by their parent's body heat, brush-turkeys bury their eggs in a mound of decomposing leaves that provides all of the heat for incubation. Only the male builds and attends the mound. Leaves are continuously added and removed to maintain a temperature between 31° C and 38° C, the optimal range for hatching eggs.

Only the male has a dangling yellow wattle, which is present when breeding. Males are also larger and noisier than the females. Each male tries to attract as many females as possible to his incubation mound. With regard to females, each may lay up to 30 eggs per breeding season distributed across all of the mounds in her local area. Presumably, for each egg produced, a female makes a separate decision about which male will fertilize the ovum (egg).

male
(length 70 cm)

Alectura lathami

Territoriality

A male brush-turkey not only defends an area surrounding his mound, but he also defends areas where food is abundant. In fact, in local parks where people gather to have picnics, it is common to find male brush-turkeys rushing and chasing one another. In the breeding season, prominent features include a long yellow wattle hanging from the base of a vividly red neck. Also, a male can vary the length of his wattle to fit prevailing circumstances. For instance, when he passes through a neighbor's territory, he has the option of retracting the wattle to minimize attracting attention, but when on home turf, an extended wattle signifies to other brush-turkeys that he is willing to defend his domain. By inflating it with air, a male's wattle is also used to produce low-

8

frequency booms, the function of which is to advertise his presence. Early in the breeding season, males spend a lot of time building and maintaining their mound and frequently stopping to *boom* to attract females.

Indeed, to have any chance of mating, a male has to encourage females to ascend his mound, which is where most copulations take place. Most visits by females, however, are only casual affairs to check whether a male is present and the suitability of his mound to receive her eggs. In these instances, a male often performs a *flattened display* by spreading his wings and tail as he lies prostrate on top of the mound. This is not to say, however, that access to a male's mound is free. If a female wants to stay for any length of time—perhaps to check the temperature of the mound, using heat sensors in her beak—she must appease the male by copulating with him; forced copulations also occur.

When ready to lay, a female first digs a hole in the mound, lays an egg in the hole, and then covers the egg. During this process the male offers no assistance. Instead, he stalks around the top of the mound and pecks at small objects, a behavior that redirects aggression away from the female. Though he wants a female to lay an egg, he does not necessarily want her to linger excessively since his primary objective is to attract as many females to his mound as possible. Needless to say, when a female is present on a mound, other females will keep their distance.

male ↑   female →

9

Visual displays of the male Brush-turkey

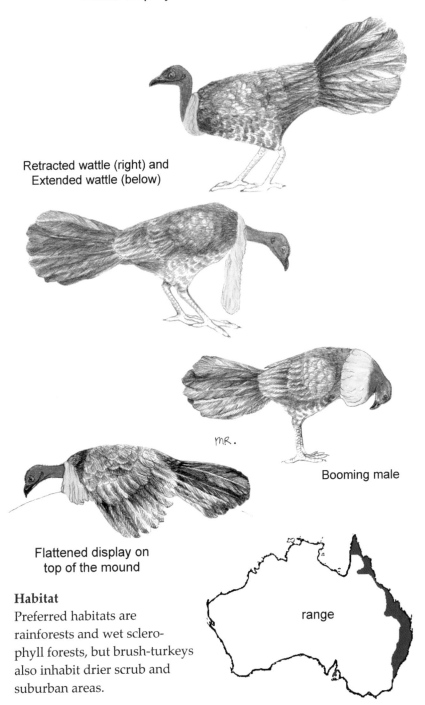

Retracted wattle (right) and
Extended wattle (below)

Booming male

Flattened display on
top of the mound

Habitat
Preferred habitats are
rainforests and wet sclerophyll forests, but brush-turkeys
also inhabit drier scrub and
suburban areas.

range

Little Penguin

(length 32-34 cm)

Courtesy of Scancolour Pty Ltd

Eudyptula minor

A group of Little Penguins leaving the water at
Phillip Island Penguin Reserve.

Breeding

Little Penguins breed in colonies and nest in burrows, caves, or under
shrubs. Although long-term bonds are common among Little Penguins,
so too is divorce, with separation rates reaching almost 50 percent in
some years. Little Penguins lay their eggs over a three month period.

When the eggs (usually two) hatch, one parent stays with the chicks
while the other leaves to catch fish. Parents continue to guard their
chicks until they can regulate their own body temperature and venture
outside the burrows. Adults spend most daylight hours in the water
fishing. They return to their burrows to feed the chicks after dusk. At
sunset, Little Penguins emerge from the sea in groups of five to 40
birds—these groups are called rafts. As they swim toward land, the
birds call to each other, giving sharp, *huck, huck* notes. When crossing
the beach to reach their nest burrows, Little Penguins are vulnerable to
predators such as foxes, cats, and dogs.

On land, mated penguins greet each other with **half-trumpet songs,**
described as "a weird note produced by alternate inspirative and
expirative moans." In the safety of the dune vegetation, the birds
frequently stop to preen and rest before proceeding to the entrance of

their nest burrow. Penguins apparently sleep for only a few minutes at a time; when walking from the beach to their nests, they often stop every 20 to 50 meters to rest and take a short nap.

The intensity of calling in the colony increases throughout the night as the adults emit *full-trumpet songs* that start on a low note and gradually ascend in pitch and volume to a harsh, almost hysterical intensity. The singing continues until sunrise when some of the adults head out to sea, while others retreat into their burrows to attend their eggs or chicks. During the day the colony is silent.

Habitat

Little Penguins breed in colonies along the coast of southeastern Australia including Tasmania (also occurs in New Zealand). They inhabit the coastal waters out to the continental shelf.

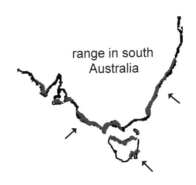

range in south Australia

Australian Pelican

(length 160-180 cm)

GRH

Pelecanus conspicillatus

Breeding

Pelicans breed colonially on sand spits and islands. Both sexes help to build the nest with males traveling the greatest distances to retrieve material. A ritualized *heads-up* display is performed when a male offers nesting material to the female.

Up to three eggs are laid over a period of several days. During periods of food scarcity, the chick that hatches first often attacks and kills its youngest sibling — siblicide may help ensure the survival of at least one of the chicks. Chicks greater than 25 days old typically join *crèches* (shown below) as they wait to be fed by their parents. After being fed, young chicks often convulse on the ground for short periods (the reason for this behavior is unknown).

Foraging

Pelicans feed on fish and crustaceans and also scavenge scraps left by humans. When fishing cooperatively, pelicans form a circle and probe within the center, thrusting their heads into the water. To minimize glare, pelicans sometimes shade the water with their wings, which seems to attract prey. They also steal prey from other birds, such as cormorants.

SNL

Crèche of pelican chicks

During long flights, pelicans fly either in a straight line or in a chevron. When flying long distances, they use rising thermals to gain altitude before soaring off toward their destination.

Pouch swinging: performed by both sexes during courtship

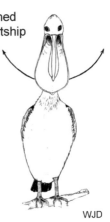

WJD

WJD

Gaping: As either sex faces an opponent, the mandibles are opened into a wide gape, indicating increased arousal.

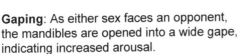

WJD

Pointing: This is performed by both male and female in response to a mild threat. With the bill shut, a bird stretches its neck toward its opponent.

Crouch-bow: performed by males before copulation

WJD

WJD

Strutting: A style of walking exhibited by a courting pair. The female takes the lead and performs a goose step-like gait as she approaches the nest. The body is held erect and the neck is slightly arched backwards while the bill rests on the neck and the wings are slightly open.

Pouch rippling (left): When courting, a male sharply claps his mandibles together to produce a ripple effect along the pouch. This is often followed with a clapping sound.

WJD

Bow and heads-up (right): To the right is shown a greeting performed when either sex approaches the nest site or each other. When bowing, the neck is arched so that the bill points down. In a heads-up display the closed bill is raised above the horizontal.

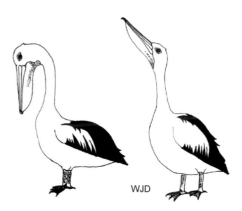

WJD

Visual display

The courtship displays of Australian Pelicans have a distinctive pattern. Initially, several males follow a female until only one male remains. The couple then performs a series of displays. While courting a female, a male sometimes throws and catches assorted items. Both sexes *strut* as they walk to the nest site. Before mating, the male *crouch-bows*, and while copulating, he arches his neck over the female's head.

Vocalizations and nonvocal sounds

Threat call: Throaty *orrh-orrh-orrhas* and *thu-thu-thus* and *ah-ah-ahhas* are given when pointing and thrusting.

Contact call: Soft *uh-uh-uhs* and *oh-oh-ohs* are given when greeting and tending chicks.

Rattle: A rattle is produced when the upper and lower mandibles are clapped together; associated with pouch rippling.

Bill-snap: A sharp clap produced when the bill is forcefully shut; associated with thrusting.

range

Darter

Breeding

Darters breed near lakes, reservoirs, swamps, and estuaries. Nests are built in trees, which typically overhang water, and occasionally within the colonies of other birds, such as cormorants, herons, and spoonbills.

(length 90 cm)

Gail R. Hill 2003

Anhinga melanogaster

Males perform a series of courtship displays including *wing-waving* which shows off the silver highlights of the wing's primary feathers. Other courtship displays include: *pointing, mutual preening, neck-rubbing,* and *bill-nibbling.* After a pairbond is formed, the couple cross their necks and adopt a stiff pointing pose. The *snap-bow* display is typically performed at the nest.

Adults use their large, webbed feet not only for underwater propulsion, but also for incubating the eggs. Both parents feed nestlings by regurgitating predigested fish. Older chicks may insert their bill into their parents' throats to obtain food. Siblings vigorously fight over food and frequently the youngest chick dies, presumably due to starvation. Nestlings typically fledge six weeks after hatching.

Foraging

Darters have absorbent body feathers and dense bones. Both features contribute to the Darter's neutral buoyancy while swimming. Additionally, they may drink water before diving to gain extra ballast. In fact, they often swim low in the water with only the head showing above the surface. Darters spear fish while floating on the surface and while chasing them underwater, at which time they paddle with their feet and steer with partially opened wings. Fish are speared by retracting the neck into a distinctive *S-shape* followed by thrusting the head rapidly forward. Reversed serrations on its bill helps a darter maintain its grip on a fish as the prey is maneuvered into position to be swallowed whole.

Darters use a number of strategies to find and catch fish. One approach is sinking slowly, which helps save energy. When completely submerged, they may spread their wings to lure fish into the shade, a behavior called *canopy-feeding*, but in this case it is performed underwater.

Out of water, Darters spend a lot of time preening and holding their wings open. *Wing-spreading* not only facilitates drying of feathers but also helps warm the bird's body the display is aligned perpendicular to the sun. Since a Darter's metabolic rate is relatively low, any extra heat can assist in the digestion of fish that have been swallowed whole. While resting, body heat is conserved by

Drying wings at the nest

coiling the neck and laying the head over the back. When it rains, however, darters may perch with the head extended toward the sky, a posture that allows water to flow off the body.

Nestlings

When threatened by a predator, nestlings may flatten themselves against the floor of the nest or when extremely alarmed, they may jump out of the nest. Older chicks often *play* by tossing sticks into the air and catching them, presumably to improve their skill in handling fish.

Vocalizations

Rattle: A series of two-note clicks that can vary in duration and volume. Rattles are typically given when a bird relieves its mate at the nest.

Threat call: A rolling series of *kaah* notes; the male's calls are usually harsher and more rapid than those given by the female. Adults and juveniles also hiss when threatened.

range

Snap-bow display: When on or near the nest, a bird raises its tail and wings and points its head down while holding the neck in an *S-shaped* curve. The wings and tail are rapidly vibrated as the bird quickly snaps its bill. Usually given by males, but also performed synchronously by a pair.

Pointing: The head and neck are fully extended and held some 20° above the horizontal. The head may be swayed from side to side as the back feathers are erected and throat feathers are flattened. Pointing generally conveys a threat but is also performed during courtship.

Wing-waving: When sitting on the nest or simply perched, males alternately raise and lower their wings, displaying the pale streaks on the body's upper surface; this display is performed to attract females.

Handling prey: Darters use their bills to spear fish underwater and then bring their catch to the surface before swallowing it whole.

Hoary-headed Grebe

(length 25-30 cm)

Poliocephalus poliocephalus

Hoary-headed Grebes feed on aquatic insects and larvae, midges, and small fish. They are highly gregarious. In the nonbreeding season, they congregate in large feeding flocks while searching for food in turbid water. At night, they also roost together, forming large rafts of birds (see below).

Hoary-headed Grebes also live in groups during the breeding season. Nests, built as floating platforms and anchored by water weeds, are spaced within a meter of each other. Hoary-headed Grebes can tolerate such close quarters because they are less aggressive than territorial grebes (and less vocal). Also, more than one female may lay in the same nest.

silhouette of a social roost

WJD

Visual displays

Hoary-headed Grebes exhibit a number of displays associated with pair bonding and courtship. When *advertising* for a mate, a pair jointly swims together with the neck vertically extended and head feathers laterally compressed. At regular intervals, the head of each participant is jerked upward as the bird calls. When a male and female meet, they may perform a *crouch-dive*, with the neck retracted and wings slightly raised as they crouch low in the water. Such performances are often followed by submerging.

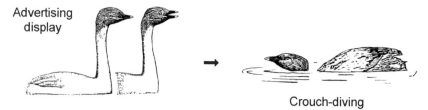

Advertising display

→

Crouch-diving

19

Though Hoary-headed Grebes are relatively peaceful, during the breeding season there are frequent confrontations over mates and nesting sites. Such skirmishes are mediated by visual displays.

Penguin dance: Birds that are highly motivated to breed show their intentions by performing *penguin dances*. In this display, a pair rapidly treads water while raising their bodies out of the water. Exposure of the white breast and ruffled body feathers enhances the display's visual effect. Following such performances, participants dive and then surface with weeds grasped in their bill—a behavior called *weed-fetching*.

Hunch-threat: When a bird is being attacked, it may retract its head and neck while pointing its bill down. Body feathers are ruffled and the wings are slightly raised.

Forward-threat: A grebe rapidly swims toward an opponent with its neck extended and held just above the water. Often the throat area is expanded by erecting the feathers.

Drawings from Fjeldsa, J. 1983.

Rearing: A bird raises its body in a highly ritualized manner and the neck is extended forward before it settles on the nest.

Head-turning: Performed while paired birds face each other. The neck is extended and head feathers ruffled. As the head is irregularly jerked from side to side, the body is turned.

Vocalizations and nonvocal sounds

Advertising call: A rolling guttural call that is given while performing an advertising display.

Duetting: Duets are given when a pair mutually performs *hunch-threat* displays.

Foot pattering: Sound of feet hitting the water. Significance unknown.

Habitat

Inhabits large open bodies of water, either estuarine, brackish, or freshwater.

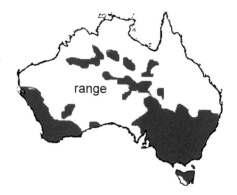

range

Australian Wood Duck

(length 44-55 cm)   male   female   WJD

Chenonetta jubata

Breeding

Australian Wood Ducks are sexually monogamous, and together a pair will search for a suitable tree hole in which to nest. The process can become quite involved since the interior of all potential holes must be thoroughly inspected. While selecting a nest site, the female calls frequently. If, on the other hand, a nest site cannot be obtained or found, a female may attempt to lay her eggs in another pair's nest, a behavior called *egg dumping*.

Although nesting material is not collected, eggs are laid on a bed of downy feathers and incubated by the female for 28–34 days. During this period, her mate remains nearby to help defend the nest from rival pairs. After hatching, the young must leave the nest to feed, and to do this they can fall a considerable distance to the ground depending upon the nest's height. When all the eggs have hatched, the hen calls from outside to encourage the whole brood to leap. Both parents tend and defend the young for nearly two months after leaving the nest.

Relative to other waterfowl, Australian Wood Ducks spend little time in water. On land, they feed on grasses, seeds, grain crops, and other green growth. During the growth period of young chicks, insects and other invertebrates are also eaten. Outside the breeding season, wood ducks travel in small flocks comprised of adults and juveniles.

Antipredator behavior

When alarmed, wood ducks become motionless and silent, erecting the mane (nape) and extending the neck vertically to get the best view. In response to the alarm calls of other species, wood ducks take flight or rush to water. To avoid detection, they can adopt a prone posture with their neck laid flat on the ground and legs tucked under their body. If a threat intensifies, parents may feign injury to lead a predator away from their young.

Display shakes: To attract a female, a male retracts his head and neck enough to touch the wings' scapulars. He then slowly extends his neck until his bill is submerged, or when out of the water, he will touch the ground before quickly raising his head to the level of his chest. While displaying, he shakes his head from side to side and often emits rumbling sounds. *Display shakes* are performed in the company of other males.

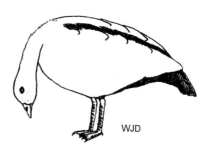

WJD

WJD

Burp display: Performed by a lone male on land or on water. The bill is held level while the neck is partially extended and the nape erected.

Turning-back-of-head: While swimming in front of a female, a male turns the back of his head from side-to-side. Often performed in a series with other displays.

WJD

WJD

Inciting (left): A female raises her chin, calls, and then spins to face the male or female opponent. When courting, the display gains intensity as a female gets closer to selecting a mate. Rarely, a male may execute inciting behavior when attending young in the company of other wood ducks.

Drink display (not shown): The bill is lowered to just above the surface. Performed before and after copulation.

Head-and-neck dip: This display usually precedes and follows copulation. Characterized by a momentary dipping of the head or head and neck underwater.

WJD

Rushing (not shown): With its neck extended parallel with the ground and the bill opened, a wood duck quickly rushes toward an intruder or predator. A hiss is emitted when rushing, and the adult may pursue the intruder for several meters.

Vocalizations

Identity call: Males emit a mild nasal *wee-ow*, while females emit a long plaintive *mew*. Given when birds reunite after a separation.

Inciting call: A nasal *wonk*, given during courtship rituals and antagonistic interactions. Most frequently given by females.

Clucks and Tooks: Emitted when feeding. A crescendo of *clucks* and *tooks* often precedes takeoffs. *Tooks* are also given by females searching for a nest hole. Also given in a variety of contexts by both sexes.

Rumble: A low pitched sound given when males perform "display shakes." Associated with courtship and could be used to attract females.

Hissing: Broadband hissing emitted during defensive and offensive aggression.

Peeps: Contact call emitted by ducklings.

range

Habitat
Since Wood Ducks prefer to forage on land, they can inhabit most bodies of water. Occupies open woodlands to grasslands including farmland and urban parks.

Dusky Moorhen

(length 35-38)

GRH

Gallinula tenebrosa

Breeding

Dusky Moorhens live along freshwater streams, lakes, and ponds across eastern Australia. In northern regions, moorhens are territorial year-round; in the south (e.g., Canberra), they are likely to abandon breeding territories in autumn to join over-wintering flocks in rich feeding areas.

Some Dusky Moorhens nest communally with breeding groups containing up to five males and two females. All members of the group are breeders, and the females lay their eggs in one nest. Weeks before laying, one or more false nests are typically built. The actual "egg nest" is usually concealed in tall vegetation. A nest that is used by more than one female may contain over a dozen eggs, since a female is capable of laying seven eggs. Both sexes incubate the eggs. After hatching, a nursery nest is often built, usually in an area surrounded by standing water, rather than hidden among vegetation. The redheaded chicks are tended by all adults in the group for three to four weeks.

Territoriality

Adult moorhens (both sexes) have a scarlet-orange frontal shield and a red bill tipped with yellow; in contrast, juveniles have a black bill and a small black shield. In some locations, during the winter months, the bill of females and younger males may fade to black or greenish.

The brilliantly colored shield is a beacon to other moorhens. From a distance, a territorial bird will signal to an intruder by raising its head and neck and staring at its rival. If the confrontation escalates, opponents approach each other with lowered head (see illustration). Such showdowns usually occur at the boundary of neighboring territories. Fights often break out with opponents biting and kicking one another.

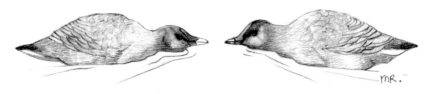

Two Dusky Moorhens confront each other.

It is unclear if any one quality of the shield influences the outcome of confrontations, whereas ownership of a territory and large body size are certainly influential. Early in the season, when territories are being established, and late in the season, when moorhens join winter flocks, the size of the shield and the intensity of its color convey the wearer's social status.

When a moorhen is threatened by a member of its own group, it is more likely to perform a *meeting display*, arching its neck and pointing its bill downward with its wings raised and tail lowered. Typically, the aggressor will respond similarly before breaking off the challenge.

Tail-flicking: With an upward flick, the tail is fanned, exposing the white undertail feathers. The tail is usually lowered slowly between flicks.

Pre-copulatory display: To invite males to mate with her, a female runs or swims in front of them with her neck extended and tail lowered and fanned.

26

Distraction display:
To distract a predator, a moorhen rides low in the water and beats its wings on the surface.

During a fight, rivals bite and kick each other as they jump and flap in mid-air.

Post-copulatory display:
After copulating, the male slowly walks away with his head lowered, tail raised, and wings arched above his back.

Dusky Moorhens build several false nests before the final egg-nest. Once the eggs have hatched, nursery nests are built to accommodate the young hatchlings.

Predator defense

An adult's response to danger is highly variable, depending upon the identity of the predator (i.e., raptor, snake, or cat) and the proximity of the threat (i.e., near vs. far). A short staccato contact call is used by adults to encourage chicks to seek cover. If a predator approaches, a series of *distraction displays* is performed, accompanied by strident alarm calls. As a last resort, multiple members in the group may simultaneously attack the predator. Despite their parents' vigilance, most chicks are lost to predators.

Vocalizations

Territorial call: raucous crowing, often returned by neighboring birds. Usually the first call given in the morning but also given throughout the day.

Click: A contact call emitted by adults to summon their offspring. Chicks respond with descending whistles and move toward the sender.

Staccato: A contact call given to warn the chicks of danger. Chicks respond by whistling and moving to cover.

range

Alarm call: A general alarm, causes moorhens to squawk and emit loud harsh calls. Intensity and type of call vary with context and type of predator.

Begging call: A shrill piping sound repeated by hungry chicks when an adult approaches with food.

Other rails of Australia

Black-tailed Native-hen

Coot

Purple Swamphen

Comb-crested Jacana

Breeding

For Comb-crested Jacanas, the domestic roles of the sexes are reversed. Males, which can weigh 50 percent less than females, incubate the eggs and care for the chicks. Females, being more aggressive than males, defend large territories that encompass the territories of several males, a style of reproduction called polyandry. The largest female frequently overcomes lesser rivals, and she then bonds with the loser's harem. The victor also kills the chicks of her rival.

(length 33 cm)

GRH

Irediparra gallinacea

Courtship involves noisy charges by the female toward a male, alert posturing and calling, and *bow displays*. While courting a male, a female stretches her neck and head parallel to the ground while slightly opening her wings. Males also fight among themselves and build several nests in their territories.

Though jacana chicks feed themselves within a few hours after hatching, their father is responsible for leading them to food, keeping them warm, and protecting them from danger. Males may even attempt to submerge the eggs to avoid detection from predators (jacana eggs normally float). To move chicks to a new location, the male will tuck them under his wings. He may also attempt to distract a predator by feigning injury and flying low and

A male carries chicks under his wings.

Lynda

GRH

erratically over the enemy, shrieking all the while. To avoid detection from above, flightless chicks may submerge themselves, while leaving only the tip of their bill exposed above water.

Most interestingly, jacanas can change the color of the comb. To quote Gail Hill, who photographed the birds, "... the female walked away from the nest and froze, with her head and neck extended skyward. As the male sat hunched over the nest, the color slowly faded from his comb; the usual rich red gave way to orange and then pale yellow. After approximately two minutes the red color returned. During this time the female's comb remained ruby red."

Habitat

Restricted to subtropical to tropical regions, jacanas occur near bodies of freshwater that contain floating vegetation including billabongs, lagoons, swamps, lakes, rivers, and sewage ponds.

Text by Gail R. Hill

Masked Lapwing

Breeding

Before breeding, individual Masked Lapwings form small groups called "piping parties" for short periods of time (10 to 20 minutes). Assembled on elevated ground, the group forms a loose circle in which individuals display to one another. Some birds leave the party by stepping sideways on stiff legs. Both paired and unpaired birds join the party.

For much of the year, lapwings move about in flocks; when breeding, however, pairs defend all-purpose territories. Both sexes incubate the eggs and feed the young. Lapwings nest in open pastures, sports fields, golf courses, gravel pits, etc. Eggs are laid on the ground in a depression which is lined with whatever soft material is available. If a site becomes waterlogged the birds may attempt to elevate the eggs by placing material under them.

(length 35 cm)

Vanellus miles

Once all of the eggs have hatched, usually within hours of each other, the downy chicks leave the nest with an attending adult. Chicks react to the alarm calls of their parents by lying flat on the ground with their head down and eyes open. They remain motionless until the parent gives an all clear call. Adults on the other hand, may attack an approaching predator or perform *distraction displays.* Feigning injury is also a common response to predators.

Masked Lapwings often nest in the middle of a mown lawn.

Territoriality

Mated lapwings frequently synchronize their offensive postures. When defending a joint territory, for example, a pair walks quickly toward an intruder, stopping every few meters to perform a *hunch display* or *body erect* defensive posture. If the intruder does not retreat, the pair may coordinate an attack.

Modified from HANZA Birds. 1993

Hunch display: The tail is raised as the bird bows and stares directly at its opponent. It may then charge its rival while this exaggerated posture is maintained.

Body erect defending: When holding its ground, a lapwing stands erect, extending its neck and crest while expanding its chest feathers. With its wings slightly raised, the spurs on its shoulders are exposed (right).

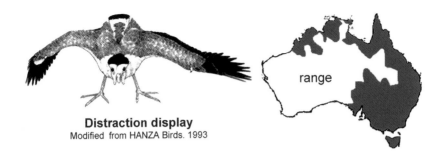

Distraction display
Modified from HANZA Birds. 1993

range

Other mid-sized long legged birds

Bush Stone-curlew

Beach Stone-curlew

photos by GRH & WJD

Australian Bustard

Buff-banded Rail

Crested Pigeon

(length 31-35 cm)

GRH

Ocyphaps lophotes

Breeding

Crested Pigeons are mostly sedentary but often will congregate in pairs or small feeding flocks ranging between three to 40 birds. In extremely arid regions, much larger flocks may congregate at water holes. Breeding can occur in a wide variety of habitats including riparian woodlands in semiarid zones, lightly wooded landscapes, farmland, roadsides, stockyards, and suburban gardens, essentially anywhere that water is available.

Both sexes share in nest building, incubation, and feeding the young. They breed year-round and typically produce multiple broods per year.

Male Crested Pigeons perform conspicuous *aerial displays* that advertise their breeding status. From a prominent perch a male rapidly ascends to 30+ meters. His vigorous flapping produces distinctive clapping and whistling sounds. With outstretched wings, he descends in a semicircular trajectory, until eventually landing on a new perch. Although aerial displays may be performed year-round, they are generally associated with nesting.

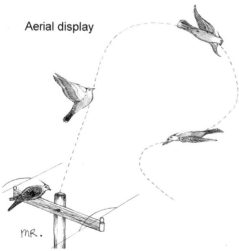

Aerial display

mR.

Crested Pigeons have a unique style of flight that includes bursts of alternating flapping and gliding. The vigorous flapping produces a whistling sound. When a bird lands, the tail is momentarily flicked over its back.

35

When courting a female, the male repeatedly performs a spectacular display by lowering his head and fanning his tail. During particularly intense performances (see right), the wings are spread open so that the primaries touch the tips of the tail feathers. When viewed from the front, the bird's wings and tail feathers form an oval, a posture that fully exhibits the male's colorful wing feathers.

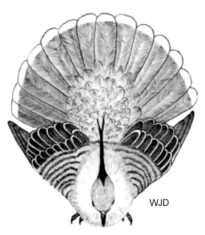

WJD

While performing a *bow display* (above), the male coos and dances, alternately raising and lowering each foot in turn.

Bill-fencing: After copulating, the male and female often gently fence with their bills.

mR.

Wing-flicking: When threatened by a rival, a Crested Pigeon rapidly flicks open one of its wings.

mR.

Vocalizations

Most vocalizations of Crested Pigeons are derived from cooing.

Advertising call: Soft, plaintive *coo-u* or *u-ru* emitted by males from elevated perches. Used to attract and stimulate females.

Display call: Short, grunting *coo* that is audible over short distances and uttered during a bow display. Females have also been observed emitting "knocking noises in the throat."

Nest call: A soft, repeated *ooroo* emitted on the nest.

Contact call: A querulous *coo-oo*.

Alarm call: A *wee wee* call.

Nonvocal sounds

Wing-whistling: Whistling sounds are produced by the outer primaries during flight. Wing-produced sounds could serve as a warning signal and, during aerial displays, as a means to attract attention.

Wing-clapping: The sound produced during vigorous takeoffs. Often produced during the initial phase of an aerial display.

Bill-clicking: When threatened, an individual may lower its crest and bill and produce clicking sounds with its bill.

Habitat

Common in suburban areas to lightly wooded grasslands. In drier areas, Crested Pigeons are concentrated near bodies of water since the birds must drink daily.

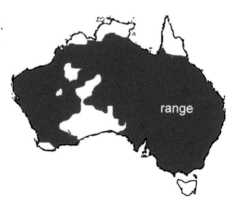

Galah

Breeding

Galahs form lasting pairbonds and both parents participate in brooding and feeding young. Nests are usually built in tree hollows, but Galahs also breed in caves and cliff holes. Soon after leaving the nest, young Galahs join a communal crèche where they wait for their parents to arrive with food. Fledglings become independent in about six weeks, at which time they join juvenile flocks.

During their second year, juvenile groups break into small flocks that move around in search of food and water. Galahs often forage on the ground, eating seeds of grasses and other plants as well as fruits, berries, nuts, leaf buds, flowers, and insect larvae.

(length 36 cm)

GRH

Cacatua roseicapilla

Eye-wiping: Galahs often remove the bark and cambium from around the entrance of the nest and then rub the side of their beak against the scarred area. Males also wipe the sides of their face on the scar, leaving behind fine powder originating from the eye-ring.

mR.

Rain dance: Galahs rarely
bathe in standing water.
When it rains, however, they
often hang upside down from
branches with their feathers
ruffled, tail spread, and wings
open to catch rain drops.

Heraldic display: Performed
year-round, a bird perches
upright with wings spread
while emitting piercing *scree*
calls and rocking forward into
a bow. Such displays are
directed at predators and
holes in trees; used to defend the nest and frighten
animals out of tree hollows.

Play: Galahs play by hanging upside
down from perches, repeatedly flying
into willy-willies, conducting *mad
flights,* and sliding down
guy wires and blades of
spinning windmills.

Illustrations redrawn
from Rowley 1990.

Vocalizations

Chet: A common call used in a variety of contexts while resting and in flight. By varying the rate of calling, *chets* can convey mild to intense excitement. As a bird becomes more excited, the rate of calling increases. Birds might also identify one another by their *chet* calls.

Cheat: Similar sounding but lasting longer than *chet* calls. Given in bouts of two to four calls. *Cheats* announce the arrival of a pair returning to their territory. Often followed by **heraldic displays**.

Lik-lik: A two-syllable call given before taking off or before a bout of stretching. *Lik-lik* calls are used to coordinate flock activity; other birds often follow the caller's lead.

Titew: A two-syllable call given with *cheats* during long distance flights.

Kwee: A soft call emitted when parents return to the nest, before feeding offspring.

Allopreening:
Galahs preen one another year-round but most frequently during the breeding season. Bouts of mutual preening occur between the same and opposite sex.

mR.

range

Habitat
Widespread wherever there are trees available for nesting. Non-breeding birds roam about in large flocks looking for food.

Galahs raised by Pink Cockatoos

Major Mitchell (Pink) Cockatoo

In western Australia, Galahs and Major Mitchell's (or Pink) Cockatoos share the same habitat. Both species nest in eucalypt woodlands and occasionally lay eggs in the same hollow. When this occurs, the Galah adults are evicted before incubation begins. Hence, chicks from both species are raised by Major Mitchell's Cockatoos (MM). At one nest, researchers recorded the behavior of cross fostered birds to determine if the young Galahs take on the mannerisms of their foster parents. As shown in the table below, some behaviors are learned while others are innate.

Behavioral Category	Innate (I) Learned (L)	Comments on the behavior of the Galahs
Begging calls	I	produced Galah begging calls even though MM fed chicks
Alarm calls	I	produced Galah alarm calls
Contact calls	L	produced MM contact calls
Flight style	L	when in a flock of MM the young Galahs flew like MM; otherwise they flew like Galahs
Food preference	L	a Galah learned to eat food fed to it by its MM foster parents
Mate preference	L	preferred MM as a mate

Rainbow Lorikeet

Breeding

Rainbow Lorikeets breed in monogamous pairs that remain together throughout the year. Both sexes look alike and share in preparing the nest, incubating the eggs, and feeding young. The nest is built in crevices and hollow spots of eucalypt trees. While breeding, lorikeets travel in pairs; at other times they congregate in flocks that range in size from a few to hundreds of birds. Large flocks are composed of smaller family units. Lorikeets form communal roosts, usually located in tall trees, when they are not breed-

(length 28 cm)

Trichoglossus haematodus

ing. Populations of Rainbow Lorikeets in many Australian cities have increased substantially in recent times.

Rainbow Lorikeets are expressive birds with a rich array of vocal and visual displays. Over twenty visual displays have been described, but such a list is probably underestimated since individual birds often combine components from different displays in novel ways.

When sitting together, pairs of lorikeets also display cooperatively. Such performances strengthen pairbonds and convey messages to rival birds. Because partners take turns displaying to each other, cooperative displays are synonymous with vocal duets. Most cooperative displays are initiated by males and many include vocal accompaniment. Most social interactions between lorikeets involve a behavior called *eyeblazing*, where the pupils are contracted causing a conspicuous expansion of the irises.

Crouch-quivering: From a crouched posture, each bird bows its head forward, slightly fans its tail, and rapidly quivers its wings. Accompanied with muted chuckles and strident twitters.

Alternating tail-flicks: While sitting next to each other, partners alternately flick their tails rapidly upward in a coordinated fashion. The birds may be perched or walking on a branch. Both birds repeatedly emit harmonically rich calls.

Alternating head-jerks: A highly coordinated display during which participants move their heads in rapid jerks, first to one side and then to the other. Partners emit a low tone that is synchronized with the movement of the head.

Hiss-ups: Each bird quickly extends its body vertically while sticking out its tongue and hissing (only one bird shown). The wings can be raised slightly or fully extended. A partner may respond very quickly, within 200 milliseconds.

Vocalizations (short list)
Lorikeets are highly vocal. Some 26 calls have been recorded, including instances of vocal mimicry by captive birds. Many calls are strident, especially when given in flight or when birds are fighting. Other calls used between pairs are relatively soft, almost conversational in tone.

High-intensity contact call: A loud, far-reaching strident shriek. Emitted in flight to advertise a bird's position.

Low-intensity contact calls: Includes a variety of squeaks, whistles, squawks, and twitters. Used when birds are near one another.

Whistle alarm: High pitched, rapidly repeated series of *tweet-tweet-tweet*. Other alarm calls have been described as barks and screeches.

Rallying call: Strident and discordant shrieks that change in pitch. Used while fighting to call in a bird's mate.

Allopreening call: Richly harmonic, low pitched throaty sounds. Used while birds are preening each other.

Habitat
A habitat generalist in that Rainbow Lorikeets adapt well to urban terrain, eucalypt forests, and rainforests. Common in suburban parks.

range

Pheasant Coucal

(length 60-80 cm)

GRH

Centropus phasianinus

The Pheasant Coucal is Australia's only ground-cuckoo. This ungainly looking bird nests and forages on the ground where it takes insects, frogs, lizards, and the eggs and young of other birds. It usually occurs alone or in pairs during the breeding season.

When courting a female, the male crouches low to the ground as he follows her about, moving his head from side-to-side and up-and-down. Sometimes a male lifts his body and head and drags his outstretched wings along the ground. During courtship rituals he often carries food, which he offers to the female to persuade her to mate with him. Throughout the breeding season mated pairs vocally duet from elevated perches with males singing a higher pitched song than the females. Males also exclusively incubate the eggs, which are laid in a domed nest built in tall, dense clumps of grass, and males provision the young with food more often than females.

Both sexes defend the breeding territory by giving harsh calls, singing, and performing visual threat displays. When pursued by a hawk, adults may drop to the ground, spread their wings, raise their long tail, and puff out the feathers on their back, neck, and head. When threatened, nestlings excrete foul-smelling feces—different from normal feces—as they burst through the back of the nest. If caught by a predator, a young coucal also excretes smelly feces.

Vocalizations

Scale call: A series of low, rich notes that descends in pitch and then rises as the rate of delivery slows. Scale calls are used in duets.

Monotone call: Similar to scale calls but softer, with each note on the same pitch. Not used in duets and does not carry as far as a scale call.

Harsh call: A scolding, hissing sound, *keouw* or *theouw*. Given in short bursts while performing a threat display.

Percussive call: A short, harsh *ptuck* given to warn family members of danger.

Contact call: Grunts emitted while moving through dense vegetation.

Hoots: Soft hoots that convey alarm to the caller's offspring and mate.

range

Habitats
Found in a wide variety of habitats including open forest, around wetlands, and suburban parks. Not often seen in the open, however, because coucals prefer dense understory vegetation common in sugar cane plantations, farmland, dense fields of grass, and Lantana thickets.

Laughing Kookaburra

(length 45 cm)

GRH

Dacelo novaeguineae

Breeding

Kookaburras are not known for exhibiting elaborate courtship displays, though a male does feed the female when she is laying eggs. Nest cavities are dug in termite nests, earthen banks, and tree hollows. A mated pair may be assisted by up to three nonbreeding helpers that take part in feeding the current batch of young and in defending the group's territory.

Helpers are usually older offspring that have not dispersed from their natal territory. Older helpers are known to challenge their parents for breeding privileges. Furthermore, adults may kill young chicks that are unable to defend themselves.

Territoriality

When two groups meet at a shared territory boundary there is usually a confrontation. The action begins when a kookaburra flies up to and perches on a tree midway between the groups. Shortly thereafter, it flies back to its own group amidst a clamor of songs and cackles from both families. The bird's retreat is often followed by one of its neighbors flying up to the boundary. Called a *trapeze display*, this scene can be repeated several times before the birds tire and fly off.

Near the peak of breeding, kookaburras become highly aggressive, causing neighbors to fly deep into their respective territories. When this happens, a resident bird typically gives chase, causing the intruder to double back. Soon after the two birds cross over their shared boundary, the pursuer becomes the intruder and is attacked. As the action unfolds, the birds fly in a figure-eight pattern called a *circle flight*. Eventually, the emotional intensity wanes and each group retreats.

Kookaburra visual displays are subtle, but nonetheless effective at close range. The *stick display* and *bill-pointing* are used in social interactions. During struggles within a fam-

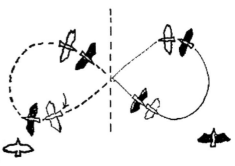

Circle flight

ily, kookaburras often grab each other's beak followed by twisting, pushing, and pulling at each other. When conflicts escalate, kookaburras resort to hitting each other with open wings, a behavior called *wing-boxing*. Frightened birds usually fluff up their feathers to increase their apparent size.

Stick displays (left) and bill-pointing (right) are used
to signal submission and mild assertiveness, respectively.

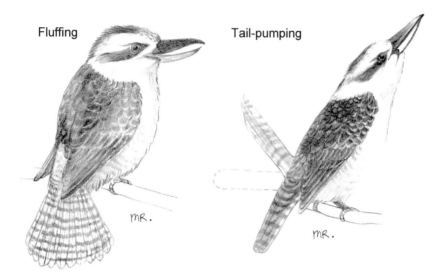

Fluffing   Tail-pumping

mr.

mr.

A frightened kookaburra (left) puffs out its feathers;
rigorous tail-pumping (right) signals agitation.

Vocalizations

Both sexes squawk, and juveniles give begging calls to solicit food from adults. Many of the kookaburra's vocal and visual displays have evolved to mediate confrontations among family members as well as between rival groups. The *full song* is sung primarily during the dawn chorus to advertise ownership of a territory and at dusk as individuals arrive at the evening roost.

Seldom is a *full song* given during the day. If a helper sings, one of the resident breeders, usually the male, immediately flies toward it and starts to aggressively spar. A subordinate bird's song is apparently taken as a direct challenge. When either sex sings early in the breeding season, its mate usually flies toward it and begins singing (i.e., a pair duets with each other). In this context, songs could be used for mate guarding.

A *full song* is comprised of distinctive sounds (e.g., *kooaa, cackle, rolling, laugh, gogo,* and *gurgle* calls). Several of these components, however, can be given alone. For example, when either the *cackle* or *laugh* is emitted separately, the caller is likely to have aggressive intentions, while the *kooaa* call is associated with fear.

(Continued on page 51)

49

Small kingfishers

Forest Kingfisher

Azure Kingfisher

Red-backed Kingfisher

Sacred Kingfisher

Dawn Chorus: Full song is given before sunrise. Used to advertise territory ownership and to guard one's mate.

Song: Full-throated, boisterous laughs are made up of multiple components. Given primarily before dawn or roosting at dusk.

Laugh: *Ha ha ha* portion of song, may be given separately and is associated with aggressive behavior.

Rolling: Fast repetition of identical syllables lasting several seconds, often followed by a full song. Associated with elevated arousal.

Gurgle: A low pitched, rolling call with long pauses between syllables. Thought to be restricted to females.

Gogo: Loud distinct "*gogo*" or "*who who.*" Thought to be restricted to males.

Squawk: Low, hoarse sound given before breeding and rarely after chicks have fledged. There are soft and loud versions of this call.

Begging: A low hissing and sometimes gravelly call. Used to solicit food.

range

Habitat
Laughing Kookaburras occur within its range wherever there are trees.

Albert's Lyrebird

Breeding

The Albert's Lyrebird is one of two species of lyrebird in Australia. It inhabits the subtropical forests of southeastern Queensland and north-eastern New South Wales. May to August a male spends time displaying from a number of sites within his territory, an area that can cover several hectares. Breeding territories are often clumped along ridges where the male's song can be broadcast over the longest possible distance.

(length 80-90 cm)

Menura alberti

The female's movements are not confined to the territory of one male, and after mating she alone builds the nest, incubates the single egg, and feeds the young.

Visual displays

The spectacular visual and vocal displays of male lyrebirds are performed at specific spots called display courts or platforms which are made of crisscrossing vines laid loosely on the ground. From this slightly elevated platform, a male can monitor his immediate surroundings without being seen. When singing, a male grasps a vine with one foot to "tap" out the beat in time with his song.

During a complete performance, the male spreads and then drapes his tail over his back to form a dome enclosing his body and head. With his tail inverted, he reveals a rufous patch of feathers normally

Display court

52

hidden under the tail. As a female approaches a male's platform, he increases the intensity of his performance.

Vocalizations
Mature males establish and maintain a territory by singing loud *territorial songs* that last five to ten seconds. Since lyrebirds are good vocal mimics (i.e., they learn to sing), males in different geographical regions sing different songs.

Once a territory has been established, a male starts to sing a softer *sequential song*, especially when a visiting female shows interest in mating. *Sequential songs* contain a number of sounds copied from other species and external sources. The sounds are woven together to produce a repetitive sequence. When singing a sequential song, males cycle through the entire sequence before repeating it. A single performance can last up to 50 seconds. That all males within a geographical area sing the same sequence of songs, known as the "albertcycle," indicates that they learned what to sing by listening to older males.

Gronking songs include a combination of very loud *gronks* interspersed with a series of soft, nonmusical notes (normally two pitches) that are performed in a precise rhythmic manner. All males in an area usually sing the same songs.

Females are mostly silent, usually giving only alarm calls and soft notes to communicate with chicks. Sometimes, however, females mimic other species and even utter the local version of the albertcycle. It is unknown how often females sing or for what purpose.

Habitat
Very limited distribution with populations restricted to subtropical rainforests and wet sclerophyll forests that have a rainforest understory.

range

White-throated Treecreeper

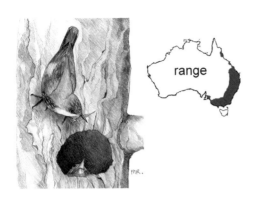

(length 13-15 cm)

DS

Cormobates leucophaeus

White-throated Treecreepers establish breeding territories in a variety of forest habitats. They hunt for bark-dwelling ants and other insects on trunks of trees and shrubs. The tail is not used in climbing. Females build a cup shaped nest in a tree hollow and incubate the eggs without a male's help. Both sexes feed the young. Females also exhibit *nest-wiping* by holding an insect, a shred of snake-skin, or a feather in the bill as they wipe around the nest entrance and inside the nest. The odor of these materials may mask the scent of the nest from predators. Threatened birds fan their tail while raising their crown feathers. Clicking sounds, produced by rapidly opening and closing the tail, can be heard during territorial confrontations and disputes between pairs.

Vocalizations

Song: A loud, long sequence of calls emitted by males.

Piping: A rollicking, sharp, high pitched *pee-pee-pee*. Used by males throughout the year to attract females.

Crescendo: Used by males to attract a mate; commonly given at the beginning of the breeding cycle and early in the morning. While calling, a male may quiver his wings and cock his tail.

Trill: A rapid, tremulous trill, *woo-woo-woo*, emitted by a male when bringing food to his mate.

Squeal: Sustained high pitched notes with an upward inflection. Heard during chases between rivals and chases leading to copulations.

Superb Fairy-wren

female

male

(length 13-15 cm)

range

GRH

GRH

WJD

Malurus cyaneus

Breeding

Mature males are strikingly beautiful, but it is doubtful that they use their good looks to impress the resident female or to repel rivals. In fact, a male that is fortunate enough to possess his own territory rarely displays to his mate, and he often shares his territory with unrelated males. Instead, fully adorned blue males fly to adjacent territories to perform eye catching displays for extra females. While away though, they leave their own mate unguarded.

Large flocks of up to one hundred Superb Fairy-wrens may forage together during winter. Come spring, each family (breeding pair plus one or more sons of the female) stakes out a territory with the larger families controlling the best real estate. In some areas, due to high mortality of evicted females, there are often more males than females. Combined, the scarcity of females and breeding habitat prevents some young males from breeding. When this happens, a male's best option might be to remain on his natal territory and help his mother raise siblings. A daughter, on the other hand, is forcibly evicted by her mother before the start of the next breeding season.

Although fairy-wrens maintain long-term social bonds, promiscuity is common. In fact, 76 percent of a female's eggs are fertilized by suitors other than the resident male, and males persistently try to attract the attention of other females. The presence of helpers at the nest further promotes this unique mating behavior of fairy-wrens. Though a female copulates with the resident male, her dependence on him to help raise the chicks decreases if she has help in feeding her current clutch of young. This independence gives her more options regarding which males will sire her future clutch of eggs. Before she consents to mate with an extra-group male, however, he must demonstrate his genetic worthiness. He does this by visiting her often and performing elaborate, eye-catching displays.

Visual displays

The simplest display of a courting male is the *face-fan* (right). To further enhance his appearance, he often carries a colorful flower petal when visiting females. Superb Fairy-wrens prefer yellow flower petals, whereas other species of fairy-wrens may select red, blue, or yellow (and berries may also be carried). While carrying a petal, the male follows the female while twisting his body from side-to-side and fanning his sky blue face feathers. Females often feign disinterest or emit an alarm call to elicit retaliation by the resident male (if he is present).

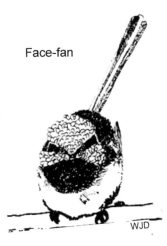

Face-fan

WJD

When evicted, intruding males perform a *seahorse display* in which the body is held in a near vertical posture, the tail is pointed downward and the male's blue and black feathers are erected to maximum effect. Such antics tell the female where the male can be found. Although a female will not mate with an extra-group male on her own territory, if she likes what she saw she will visit him later on his own turf. Visits by extra-group males occur throughout the year, but the number of visits to a particular female increases a week before she starts to lay, during the period when she is fertile.

Carrying a flower petal

Seahorse display

Response to predators

Conspicuous postures are performed by fairy-wrens to divert predators away from the nest. One in particular, called the *rodent-run*, resembles a mouse running along the ground.

While performing a *rodent-run*, harsh sounding alarm calls are usually uttered. Occasionally, a bird exhibits a different version of this display by running along a branch or climbing up and down a stem of grass. Alternatively, a fairy-wren may simply remain stationary while moving its head back and forth.

Rodent-runs are performed only after the eggs have hatched (two to three days old), and then the alarm call of nestlings is necessary to evoke this behavior. Most curiously, while displaying and running, a fairy-wren may pick up a flower petal as if trying to become as conspicuous as possible. When a nest is disturbed, an adult may hit the ground like a falling stone and start performing the *rodent-run*. Finally, other birds such as fantails and thornbills are attracted to the fairy-wren's alarm calls and often help to mob the predator.

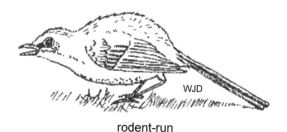

rodent-run

Striated Pardalote

Striated Pardalotes inhabit eucalypt forests, woodlands, and scrublands. They are more easily heard than seen since they spend most of their time foraging high in trees, often in flocks up to one hundred birds. They glean insects and spiders off foliage and harvest the sugary excretion of insects called lerps (psyllids) that feed on eucalyptus leaves. In some regions Pardalotes remain in the same area year-round whereas pardalotes in other regions seasonally migrate between breeding and nonbreeding locations.

(length 8-10 cm)

WJD

Pardalotus striatus

Striated Pardalotes usually breed in pairs with a few reports of cooperative breeding (trios). A loosely constructed dome shaped nest is built in a tree cavity or at the end of an earthen tunnel. Both sexes help excavate nest tunnels.

mʀ.

Wing-and-tail display

Not much is known about the visual displays of pardalotes. They do perform a conspicuous *wing-and-tail* display near nest holes or, occasionally, within a group of birds near nest sites. While displaying, the wings and tail are spread and held open and fanned for several minutes at a time. The function of *wing-and-tail* displays is unclear.

Vocalizations

Song: A typical song consists of two to seven short notes (whistles). Singing is largely confined to the breeding season. Distinctive songs are sung by birds breeding at different geographical locations—Striated Pardalotes include a number of different races across the country, each with a slightly different song.

Contact calls: Contact calls are most often given in flight and comprised of six to 13 down-slurred notes.

Habitat
Widespread across Australia, occurring everywhere there are trees and shrubs; favors eucalypt forests.

range

Noisy Miner

Territoriality

Noisy Miners form complex social ties with one another. Although some males may defend individual territories, the majority of males band together to form loose coalitions that, in turn, become part of a tightly knit network called a *coterie*. Combined, all coteries in the neighborhood form a colony.

In contrast, a female's activity space may overlap with one or more male coteries, but not with real estate controlled by another female. Also of interest is the fact that Noisy Miners defend their territories from intrusion from a variety of other species of birds, particularly species that consume similar food.

(length 24-28 cm)

Manorina melanocephala

Breeding

During the breeding season, females fight over the best habitat. Once established, a female may copulate with multiple males; yet more than 95 percent of her young are typically sired by only one male. Females build the nest with occasional help from males. In fact, to attract the attention of visiting males, females advertise the location of the nest by performing a conspicuous *heads-up flight* display.

In this cooperatively breeding species, several males usually help feed a female's offspring. Although some males may visit the nests of several females, most helpers are sons of the female that they are assisting. Females are rarely helpers because they are driven out of their natal area by resident females.

Social ties among Noisy Miners are maintained by a rich array of vocal and visual displays. The most conspicuous display is the *corroboree* that involves multiple birds converging at a single location and simultaneously displaying to one another.

Heads-up flight: As the female flies, her head is held up and back while the body and tail are held in a vertical position and the feet dangle. Males may give this display when approaching a nest to feed nestlings.

Corroboree: A striking social display lasts nine to 30 seconds and involves a number of birds. Participants simultaneously display to one another. Many submissive displays are given during a corroboree.

Pointing: A closed bill is pointed at another bird while the neck is outstretched; feathers on the head, neck, and body are sleek while the wings are held closely against the sides of the body. After several seconds, pointing often leads to *wing-waving*, especially while mobbing another miner.

Eagle posture: Wings and tail are outstretched with feathers spread in a vertical plane. A closed beak is oriented toward the recipient of the display and the wings are vibrated slightly. Used by males to signal submissiveness to a female.

Eye displays: Noisy Miners can change the size of their brilliant yellow eye patch by manipulating the surrounding feathers. The eye patch probably functions as a coverable badge.

mR.

Tall and low: A dominant miner may face its companion and extend its height by stretching its legs and neck. The bill is closed and pointed downward. In response, the subordinate bird turns its face away as it crouches and fluffs its feathers.

mR.

Vocalizations

Singing: During the dawn chorus, members of the colony sing for periods of four to 13 minutes. Some populations of miners sing more than others.

Duetting: When breeding, duetting is common between males and females, with a female chuckling in response to the *short-flight* call of a male.

Short-flight call: Rhythmic calls, usually consisting of four syllables. Given by males while conducting short flights; the quality of the calls varies regionally.

Chuckling: A series of chips given by females in response to a male's short-flight call.

Scolding: A harsh broadband sound given when birds are mobbing or harassing an intruder or predator— e.g., Australian Magpie, Magpie-lark, Pacific Baza, or currawong.

Yammering: A rhythmic call emitted when birds come together in an excited state.

Contact notes: A short series of ascending whistles emitted under a variety of circumstances, usually while preening or foraging.

Alarm call: Miners have two, maybe three, distinctive alarm calls comprised of short whistles that may descend, ascend, or remain constant in pitch. One type of call is given in response to aerial predators; a second type is given in response to terrestrial predators such as snakes and Goannas (large land lizards). Noisy Miners also emit alarm calls in response to alarms of neighboring miners, provoking the whole colony into a bout of calling. Hence, calls are often given well before a predator comes into sight.

All clear: A short call emitted when the danger has passed.

Food call: A call given when a substantial food source is discovered. Upon hearing the food call, other miners respond by approaching the sender.

Habitat
Resides in open woodlands and forests. Very common within the species range wherever humans have modified the habitat.

range

63

All four miners

Bell Miner

Yellow-throated Miner

Noisy Miners

Black-eared Miner

New Holland Honeyeater

(length 16-19 cm)

GRH

Phylidonyris noveahollandiae

Territoriality

New Holland Honeyeaters are pugnacious. They often chase and harass other birds that threaten to raid the local nectar supply. When nectar production is low, nonbreeding birds join flocks that roam over extended areas looking for new food resources. In nectar-rich areas, however, pairs of honeyeaters jointly defend breeding territories or at least protect a few patches of flowering shrubs and trees that supply them with food.

New Holland Honeyeaters occur in loose colonies in which resident birds, although territorial, are tolerant of other members of the group. Strangers, on the other hand, are attacked.

Visual displays

When New Holland Honeyeaters approach one another, they exchange visual and auditory displays that reveal their degree of dominance or submissiveness.

Two honeyeaters perform *low postures*. The bird on the left is enhancing the intensity of its display by fluttering its wings.

The *high posture* given by the bird on the right signals its higher rank. The bird on the left responds by giving a *medium posture* and by flattening its facial plumes.

Individuals typically adopt a *low posture* when a more dominant bird approaches. Submissive displays are enhanced by fanning the tail and fluttering wings that are partially outstretched. To signal an increase in confidence or assertiveness, a bird raises its body to a *medium posture* and adjusts the angle of its beak toward its opponent. When seriously confronted a dominant bird may adopt a *high posture* by raising its body, extending its neck and angling its head so as to look down on its opponent.

The extent to which the ear plumes are erected also signals how excited or aggressive a bird is likely to become. The ear plumes of submissive birds are invariably flattened against the head, whereas ear plumes are erected when a bird adopts a *high posture*. In New Holland Honeyeaters, erected ear plumes signal dominant status. In fact, if a bird's plumes are removed, it cannot maintain its status within a group.

At the start of the breeding season, when territories are first established, the bouts of displays become more elaborate and prolonged. The sight or sound of two birds interacting will often prompt other honeyeaters to join in on the action. The phenomenon of numerous honeyeaters (three to 20 birds) displaying to one another is called a *corroboree*.

A *song flight* is a male's most conspicuous display and is performed near a nest site, typically over a clear area. A flight may extend 50 meters before the male breaks off and glides back to where the flight began; a song is usually sung during the last half of the flight.

It is believed that *song flights* serve as an "all clear" signal, a kind of distraction display that is given for the benefit of a nesting female. When females are away from the nest, they have been observed to stop feeding and cease moving during the performance of *song flights*.

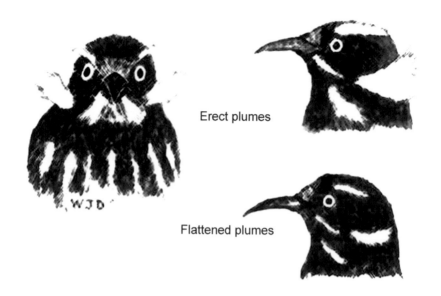

Erect plumes

Flattened plumes

Observations also indicate that females delay their approach to the nest until after their mate performs a *song flight*.

Song flights originate from elevated perches where the male is in a good position to detect potential danger, and they are often performed when a male has been disturbed by a predator. In this context, a *song flight* may alert a female to potential danger or divert a predator's attention away from the nest.

Habitat
New Holland Honeyeaters inhabit wooded areas, gardens, heaths, and forests—especially where grevilleas and banksias grow. Their cup shaped nests are made of bark and grasses that are bound together with spider silk.

range

Regent Honeyeater

Breeding

Regent Honeyeaters tend to breed in loose clusters with nests placed as close as 10 meters apart, although distances of 30 to 50 meters are more typical. Still, a breeding pair aggressively excludes all other honeyeaters from their small breeding territory, and solitary nesting pairs are common in some areas.

Males facilitate nest site selection by giving a *wing* and *tail shivering* display at potential nest sites. Once a site has been accepted by his mate, usually in the fork of branches of rough-barked trees, the female starts building the nest from plant material and spider silk.

Males advertise their territories by singing *quipper-quip* or *quippa-plonk-quip*. These songs are accompanied by conspicuous *head-bobbing* and audible *bill snapping*. Singing continues until the eggs have hatched. Both sexes share in the feeding of the young.

Regent Honeyeaters learn their songs and also mimic the sounds of other honeyeaters, especially larger species. Captive raised Regent Honeyeaters have, in the absence of adults, developed songs that bear little resemblance to that of wild regents, indicating that they must learn what to sing.

(length 20-23 cm)   GRH

Xanthomyza phrygia

WJD

Head-bobbing

68

Grey-crowned Babbler

(length 25-29 cm)

Pomatostomus temporalis

Breeding

Highly social Grey-crowned Babblers live in sedentary groups in semi-arid regions of Australia. They build domed nests near the tops of trees in which they roost and breed. Typically only one pair within the group breeds, although the so-called primary female may copulate with neighboring males, and two females may lay eggs in the same nest.

Group membership includes males and females of different ages, and most are offspring of the breeding pair. Occasionally an outsider joins the group. Babblers often move together in a coordinated fashion. One group strategy for crossing an open area is particularly intriguing to watch. With all members positioned at the top of a tree, one individual takes flight while calling. Soon afterwards other babblers in the group follow one by one and their arrival is spaced one to four seconds apart. Each bird follows a similar flight path, creating a procession of birds.

When threatened, babblers adopt defensive postures within a *defensive clump* (see below). At night the entire group usually roosts together within one of the domed nests. In the breeding season, the primary pair spends the night alone in the brood nest.

Defensive clumping

69

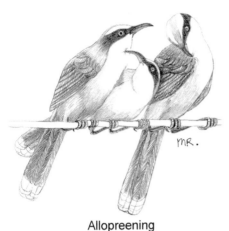

Babblers often take turns preening one another. Such *allopreening* helps to maintain group cohesion as well as to groom hard to reach feathers. A second maintenance activity is *anting*, whereby a bird lies on or near an ant nest to entice the insects to crawl over its body. When disturbed, some ants emit formic acid that can fumigate the bird's plumage.

Allopreening

Babblers *play* by pulling at each other's feathers, rolling on the ground, and grabbing and carrying items. Juveniles and occasionally adults spar by grasping each other's bills and twisting their heads and bodies.

Distraction: Distraction displays are performed in response to predators and distress calls from other babblers. The body is held low, the tail is erected and opened, the head is inclined slightly, the beak is opened, and the wings are quivered.

Courtship ritual: In an upright position with their bodies touching, tails spread, and wings closed, a courting pair will turn toward each other and stretch their necks. After a bout of bill touching and wing fluttering, the pair copulates. Other members of the group do not participate.

Tail-flicks: Performed most often after flying or hopping. By exposing the white underside of the tail, tail-flicking functions as a visual signal associated with movement.

Submissive posture: The body is held low and horizontal, with the head held lower than the bird's opponent. The tail is closed, with quivering wings held slightly open.

Huddle: A group display that may last five to 45 seconds and involves multiple babblers crowding together on a branch. Huddles are usually initiated by *chuckling* from the group's primary breeders. Other babblers joining the huddle also begin to vocalize. Primary breeders are the only birds that assume an upright posture and duet. After one or more duets, the huddle disperses. Duetting may also initiate a huddle.

Vocalizations

Babblers possess a large repertoire of calls. Only a select few vocalizations are described below.

Ook-ai: A loud, rhythmic call with syllables emitted in pairs. Given when birds are separated by more than 20 meters and when the primary pair leaves the nest. Also heard occasionally in duets.

Wee-oo: A high pitch whistle audible for over 200 meters. These calls are used by males in duets and also function as contact calls.

Flight call: A soft *chuck*, with syllables emitted in a series of dyads while flying. Given when leaving or arriving at the brood nest or during *group flights* (see sonogram). Probably indicates that the sender is about to take off or land.

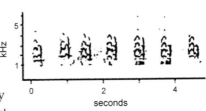

Alarm calls: There are four distinct types:

- low intensity mobbing calls comprised of a single *skak*, evoked by   general disturbances
- high pitched multi-syllable call given while in a group
- a series of harsh *skak* notes given most often when mobbing
- high pitched whistle given when a flying raptor is spotted (Note: Usually all birds in the group start calling as they fly for cover. Attacks from Magpies and Noisy Miners may evoke the raptor alarm call.)

Distress: A call comprised of short, rapidly repeated notes. Evoked when a bird is captured or handled. Group members usually approach the caller and start to perform distraction displays.

Chuckling: Loud, rapidly repeated call resembling barking. Given by members while in a *huddle*.

Duetting: Duets are performed by breeding babblers while they *huddle* with other family members. The distinct *ya-hoo* is commonly given in duets. Other sounds may also be included in a duet. Usually only the breeding pair duets, unless one member of the pair is away (e.g., female incubating) in which case another bird (usually another female in the group) may duet with the primary male.

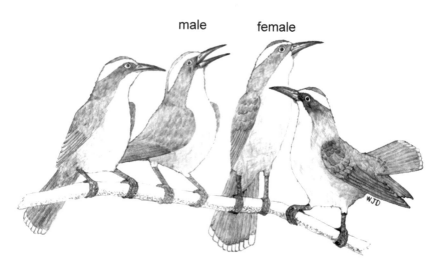

male   female

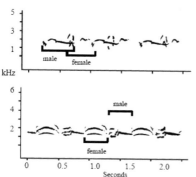

A pair of Grey-crowned Babblers (center birds) are shown duetting within a huddle. Sonograms (left) show separate duets sung by the pair. (All drawings modified from B.R. King 1980.)

Habitat

Inhabits open forest, woodland, and areas with substantial covering of shrubs. Prefers areas where there is considerable debris, leaf litter, and fallen timber for foraging.

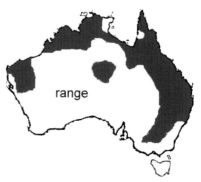

range

Willie Wagtail

Breeding

Unlike other Australian fantails, the Willie Wagtail prefers semi-open terrain, including suburban gardens, parks, and pasture land. They are nonmigratory and maintain prolonged pair-bonds. Both sexes help build an open cup nest on top of a tree branch. The nest is built primarily with spider silk.

Presumably to protect their young, Willy Wagtails often build nests near aggressive but non-predatory species such as the Magpie-lark (drawing below) or the White-winged

(length 19-21.5 cm)

Rhipidura leucophrys

Triller. Individual territories are defended year-round and jointly defended by paired birds during the breeding season. Willie Wagtails are known for their fearlessness when *mobbing* predators such as wattlebirds, magpies, crows, kookaburras, lizards, dogs, and cats.

mR.

In some parts of their range (e.g., Papua New Guinea), wagtails build multiple nests in the same tree before laying eggs in one nest. The empty nests may confuse predators; birds sometimes even sit in empty nests as if embellishing the ruse.

Foraging

Willie Wagtails sally for insects on the wing, pounce on prey from a perch, run after insects

74

on the ground, glean food off the trunks and leaves of trees, and steal food from other birds. In grazing country, Willie Wagtails are called the Shepherd's Companion because of their habit of hawking insects from the backs of sheep and cattle. Where crocodiles are common, wagtails remove morsels of food from the reptile's teeth.

Willy Wagtails often sing at night and while sitting in the nest.

While stalking insects on the ground, wagtails may partially flash open their wings and persistently wag their tail. On the ground, the tail is usually closed; in trees, however, the tail is often "fanned" open as it is in other species of fantails. Is *tail-wagging* used to maintain balance or communicate with other wagtails? The fact that tail-wagging is performed less often in bright sunlight than in the shade, suggests that it is used to startle cryptic prey out of hiding.

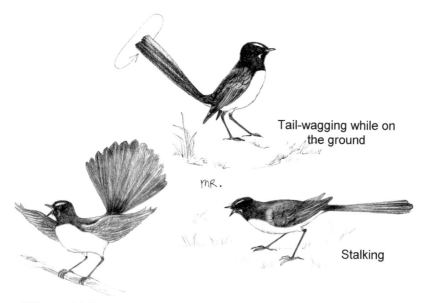

Tail-wagging while on the ground

Stalking

Wing-and-tail spreading while foraging in a tree

Visual displays

Willie Wagtails exhibit a large array of vocal and visual displays. The *dance display*, involving a male and female, is performed a month or two before breeding commences. When a pair sits on a branch, the male bobs toward the female who is crouched usually with her white eyebrows extended. The male then jumps toward and around her while singing or chattering; he may

Dance display

also swing under the branch where she is sitting. During a dance, the male's eyebrows are also raised and both birds sing. They may sit side-by-side and sing a duet before flying off together. Often the song of a lone wagtail will entice its mate to approach and also sing. This is especially common early in the breeding season.

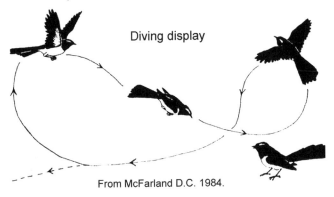

Diving display

From McFarland D.C. 1984.

A *diving display* is performed when rivals confront one another at a common territorial boundary. Expanding its white eyebrows, one bird conducts an aerial loop and dives directly at his or her opponent while singing or scolding. The bird's opponent may take its turn at diving or may flee back into its own territory. The mobbing of larger birds by Willie Wagtails closely resembles diving displays with one notable difference: wagtails performing diving displays do not make physical contact, even though birds confront one another head-on. Hence, the display has evolved from mobbing into a ritualized form of combat.

Vocalizations

Dawn (and nocturnal) song: Short, loud songs sung by both sexes usually just before dawn or on bright moonlit nights.

Song: When singing during sexual interactions, a song is quick and rattle-like. Both sexes sing in various contexts throughout the breeding season. They also sing while sitting on the nest.

Chatter: A common call that resembles the sound of wooden matches shaken in a matchbox. Given when excited, while flying, perched, and mobbing.

Alarm call: A high pitched call emitted when a predator is spotted.

Habitat

Willy Wagtails are widespread in open terrain that contains some trees; it is very common in urban habitats, but absent from rainforests. Wagtails build their nests near water courses and wetlands.

range

Other fantails

According to some authorities, there are six species of fantails in Australia. Of these, the Grey and Rufous are locally common in forested habitat in eastern Australia; the Arafura, Northern, and Mangrove Fantail occur in coastal regions in the north central parts of the country.

Grey Fantail (left) and Rufous Fantail (right) are forest dwelling species.

Satin Bowerbird

Male Satin Bowerbirds build avenue bowers that they decorate with flowers and other colorful objects. The quality of bowers and the ornaments used to decorate them are traits evaluated by females when deciding whether to mate with the owner of the bower—males do not supply females with food, a territory, or a nest site.

(length 27-33 cm)

GRH

Ptilonoryhynchus violaceus

The process of mating in this species is elaborate and multi-faceted. Though the quality of a male's bower is important, his courtship performance is also influential. When a female approaches a male's bower, he grabs one of his decorations and begins his song and dance routine. The male frantically bounces back-and-forth, fluffing his feathers and producing a wide array of bizarre sounds. In fact, males often mimic a variety of sounds that they integrate into their song repertoire. A male's performance becomes particularly intense when a female descends to the ground and enters his bower. In order to mate, a male Satin Bowerbird must not only build an attractive bower and give a convincing courtship performance, he must also prevent other males from interfering.

Both sexes of the Satin Bowerbird are versatile singers. While males sing to attract females, females sing at the nest after the eggs have hatched. Also, they are likely to mimic the sounds of potential nest predators such as ravens, butcherbirds, Goshawks, and Wedge-tailed Eagles. Unfortunately, very little is known as to why female bowerbirds sing.

mR.

Significance of the bower

Males that maintain well constructed bowers are more likely to secure the highest number of sexual solicitations from females. Hence, two strategies that maximize breeding success are maintenance of attractive bowers and demolition of bowers of rival males.

Inter-male conflict can be intense, though physical combat during confrontations is rare. Pilfering ornaments is an easy way to enhance the beauty of one's bower while undermining the sexual attractiveness of a rival.

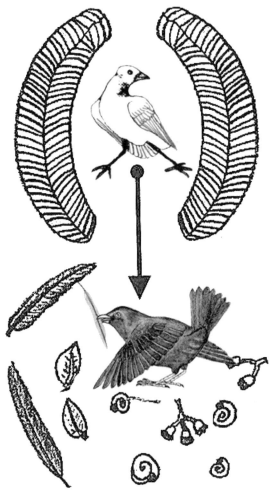

A female (above) inspects a bower that is decorated with colorful items. From within the bower's walls she can safely watch the male perform for her. Adapted from Doucet and Montgomerie, 2003.

For unknown reasons, female Satin Bowerbirds are attracted to blue ornaments. Before humans arrived on the scene, blue items included fruit, flowers, and feathers—none of which were particularly abundant. In remote areas today, for example, the average number of blue items displayed at a bower is five; whereas some bowers built near human dwellings may exhibit over a hundred blue items, including straws, hair-clips, and bottle caps.

Naturally, as the availability of blue items increases in an area, the importance of blue ornaments as status symbols would be expected to decrease as well. Perhaps this is why the rate of stealing blue ornaments is low in areas where bowerbirds live near humans as compared to populations of bowerbirds living in more remote areas. Instead, urban males choose to destroy a rival's bower, thus reducing his ability to attract females. In fact, because of rampant bower destruction, the majority of urban males are lucky to mate even once per season.

Bowers have another function: they protect females from the aggressive advances of hormonally charged males. As in many other species, courtship often becomes highly aggressive, even out of control; protected by the walls of a bower, a female can stand relatively close to a male as he jumps about and occasionally charges the bower. From within the bower, she can assess the male without risking physical harm or unwanted sexual advances. In other words, the bower empow-

Male Satin Bowerbird presents a leaf to a female.

ers the female to select her sexual partners. A male cannot ignore a female's gestures if he wants to persuade her to accept him as a mate. Female Satin Bowerbirds convey their willingness to mate by *crouching* in front of the male.

When searching for a suitor, a female may visit half a dozen bowers. When she approaches a bower, a series of predictable events unfolds. When she enters the bower, the male picks up a decoration and vocalizes; soon after, he begins to flap his wings, usually synchronized with

Intensity of crouching

Both the depth and the rate a female repeatedly crouches signal her receptivity to a male's advances. Adapted from Patricelli, et al. 2004.

loud buzzing vocalizations. Next he puffs out his feathers as he runs in front of the bower followed by the sudden opening of his wings. Inexperienced females are often startled during the buzz/wing-flapping segment of a male's display. A male can avoid scaring off the timid, young females that visit his bower by simply adjusting the intensity of his display in response to the reaction of his audience. Apparently, this is exactly what the most successful males are capable of doing.

Painting bowers

Putting the finishing touch to a bower includes coating the inside sticks with saliva or saliva mixed with charcoal, liverwort, moss, or some other plant material. What this practice means is not clear, but we do know that males have been observed painting the sticks within a rival's bower when the owner was away. If discovered in the act, the intruder is not only evicted but the owner immediately recoats the exact spot where the intruder applied his "paint." Since females have been seen "tasting" the paint applied to bowers, can we assume that the practice is associated with courtship?

Bower orientation

To maximize the visibility of the male when displaying, bowers of the Satin Bowerbird are typically oriented along a north-south axis, especially in open canopy forests. Such alignment directs light onto the display platform in front of the bower where colorful objects are on display.

Construction of a bower includes painting the inside walls with plant juices and saliva.

In rainforests, however, the closed canopy blocks much of the incoming light. Such is the case in northern Queensland where bowers of the Satin Bowerbird lack any specific compass orientation. Yet males make an effort to choose well illuminated locations to construct bowers. Instead of adopting a north-south orientation, males invariably place bowers near gaps in the forest canopy with the front entrance facing away from the light's source. This alignment maximally illuminates the platform area in front of the bower. Also, in rainforests, bowers are often built on slopes and oriented uphill, which should further enhance frontal illumination.

range

Habitat

Satin Bowerbirds inhabit wet woodlands and forests. A male usually places his bower along the edge of an opening. Birds in the Atherton Tableland region are largely rainforest inhabitants.

Females build their open cup nests out of sticks either in a tree or bush.

Other bowerbirds

Regent
Bowerbird

Female Great Bowerbird
inspects a male's bower.

Spotted Bowerbird
(left)

The Green Catbird (right)
does not build a bower.

Pied and Grey Butcherbirds

Pied Butcherbird
Cracticus nigrogularis
(length 32-36 cm)

range

Grey Butcherbird
Cracticus torquatus
(length 24-30 cm)

To handle large prey, butcherbirds impale food on sharp sticks or wedge items between forks in trees before ripping them into small pieces. The "butchering" process is aided by a small hook located at the tip of the bird's beak. Prey of butcherbirds includes insects, small birds, reptiles, and mammals.

Butcherbirds are also accomplished singers, with pairs singing duets early in the breeding season. Pied Butcherbirds sing from conspicuous locations in the canopy of trees; whereas the Grey Butcherbird generally sings at a lower height and at the outer edges of branches. As a Grey Butcherbird sings, its tail and wings open and close in sync with the rhythm of the song. Pied Butcherbirds perform magnificent aerial displays during territorial disputes.

Both species nest as pairs or in small family groups. They are unafraid of humans and are known to dive-bomb intruders that threaten the nests. Generally, butcherbirds coexist peacefully with humans.

A Grey Butcherbird wedges a beetle between two branches.

Grey Butcherbird wags its tail while singing.

White-winged Chough

Territoriality

White-winged Choughs live in tightly knit family centered "mafias," overtly harassing their neighbors and kidnapping their young. Within some groups, young birds practice deception to avoid giving up food. Much of the choughs' bizarre behavior facilitates survival in dry eucalypt woodlands where food is difficult to find.

During the winter, families roam over large areas in search of food. When rival groups come face-to-face, instead of fighting they resolve

(length 43-47 cm)

Corcorax melanorhamphos

disputes by exhibiting ritualized displays that feature tail-wagging, wing-flapping, and high pitched whistling. When one bird starts displaying, others usually follow its lead. Visual displays are enhanced by the chough's white wing patches and bright scarlet eyes that can become engorged with blood (technically, the conjunctiva surrounding the eye fills with blood).

During ritualized face-offs, adults often give full *open-wing* displays. The visual impact is enhanced by the white patches on the wings and by the bird's red eyes. Also notice the adult adopting a hunchback posture while flying back to its group. (adapted from Heinsohn 1997).

WJD

Social creatures, White-winged Choughs take dust baths together.

White-winged Choughs travel in groups to ward off attacks from other species, particularly Australian Magpies that are faster and heavier than choughs. When threatened, a lone chough emits an alarm call that quickly recruits comrades to its defense. Clumped together, the choughs utter a series of loud hisses and open their wings to exaggerate their size.

Breeding

In early spring (August–September), choughs restrict their movements to a small area surrounding the nest. Since relations between neighboring groups can become hostile, sentinels are posted near the nests. Typically, a single predator, such as a currawong, is mobbed; whereas the sight of a rival clan of choughs evokes the formation of a defensive circle around the nest to prevent its destruction. Interfering with a rival group's reproduction is an effective way to curtail the growth of a competitor's clan.

Like other birds that breed cooperatively, young White-winged Choughs stay with their parents for several years—during this period they are expected to help raise younger siblings. If many young birds survive their first year and breeding conditions remain favorable, over a period of several generations the size of a family from two parents can reach 20 birds. (In New South Wales eight birds per group is the average.)

Raising nestlings

Incubation of the eggs is the duty of the laying female. When it comes to feeding nestlings, however, all mature adults in the group are expected to contribute. Most pairs and trios (a pair and one helper) invariably fail to raise a full clutch of chicks because a small group cannot supply multiple chicks with enough food.

Dependence of juveniles

Young choughs leave the nest between 20 and 28 days after hatching. Newly fledged, brown eyed juveniles remain dependent upon the group for food for six to eight months, gradually increasing their foraging efficiency over a period of several years.

To find food (e.g., frogs, ants, centipedes, and assorted insects) choughs rake through the leaf litter with their long, curved bill. The staple diet for feeding young is subterranean beetle larvae, which requires considerable skill and effort to procure.

Dependence on hard to find prey severely limits the amount of food that a juvenile chough can harvest. Individuals not only have to learn where to dig but, equally importantly, when to stop digging if a larva is not encountered. Learning to feed oneself does not come easily to choughs, which is one reason why juveniles remain dependent upon adults for such an exceptionally long period. Understandably, most juveniles prefer to beg for a meal than forage for themselves, and freeloaders usually do better than more independent juveniles. Eventually, over a period of two years, all surviving young have gained enough experience to feed themselves. (Sexual maturity is not reached until the fifth year.)

Deception in the ranks

As previously noted, by two years of age juveniles are "expected" to help feed younger siblings, but often there is a conflict of interest as an inexperienced bird may struggle to feed itself. As a result, some young birds simply go through the motion of feeding nestlings if there is no other chough looking on to monitor its behavior.

Approximately 30 percent of the trips to the nest by two-year-old birds are false-feedings followed by preening of nestlings immediately after the helpers ate the food themselves. When researchers provided extra food to choughs, all false-feedings stopped. The "ostentatious" preening, on the other hand, seems to be an attempt to escape "punishment" from adults. Obviously, helping to feed younger siblings increases the chance that a relative will survive; it can also elevate a helper's social prestige within the group.

Kidnapping

Large groups are not easily harassed by other birds, and surplus helpers assist in raising the group's current brood. If all goes well, group size gradually increases over several years. The rate of growth can be

During display-battles, members from a rival group may approach and vigorously display to unattended chicks to persuade them to join the rival's clan.

accelerated by kidnapping chicks from neighboring groups. In one study, 20 kidnappings over four years were recorded.

Transfer of young between groups is not a violent event; conflict is highly ritualized, with members from warring groups lining up on opposing branches, often in the same tree where they loudly vocalize and vigorously wag their tails and flap their wings. Fledglings that cannot fly remain on the ground where they are vulnerable to the charms of rival birds. If not interrupted, the youngest choughs can be persuaded by displays and bouts of preening to follow their kidnapper back to the opposition's camp—very young choughs cannot recognize members of their own group. Kidnapping could be a win-win situation if a chick's chance of survival and opportunity to breed is better in its new "family."

When dynasties break up

During severe droughts, high mortality can disrupt large family groups, causing birds to form new coalitions. Within these newly formed groups, social dynamics are volatile. Whereas all chicks in a stable group are produced by one breeding pair, in the newly formed groups, multiple parentage can occur when a female mates with more than one male (polyandry) and males and females mate with multiple partners (polygynandry).

Coalitions of related individuals (most often among males) exist within the larger groups that are governed by separate dominance hierarchies. The dominant bird (usually the oldest) is the most sexually active, and his effort to breed is supported by other coalition members. Among alliances of females within the group, there is competition to lay eggs, with individuals tossing each other's eggs out of the same nest.

Australian Magpie

Breeding

In some regions, magpies breed cooperatively, with more than two adults occupying a territory. Pairs and trios are most common in southeast Queensland; whereas in Western Australia, up to 20 birds may comprise a group.

As a rule, there is one breeding pair per group. The female alone builds the nest and incubates the eggs while both parents feed the offspring. When females outnumber males, however, bigamy and trigamy may occur, and some females may resort to laying eggs in another female's nest. Nests are built in the canopy of a tree or in a location that gives the occupants full exposure to the sky.

(length 36-44 cm)
male

Gymnorhina tibicen

In some areas, young birds stay on their parents' territory for two years or join flocks of non-territorial birds when they are fully fledged. Most magpies reach sexual maturity in their second year, but few manage to breed at this time because breeding habitat is scarce.

Territoriality and flocks

Across the continent, Australian Magpies inhabit a wide variety of terrain. This is possible because social behavior of magpies is adaptable. In productive habitats, for example, pairs establish permanent territories year-round. However, this arrangement is not viable in impoverished areas; here, magpies may defend only a small area around the nest. In regions where nesting and feeding sites are separate, magpies establish "mobile" territories around feeding sites. Then there are cohorts of birds, for one reason or another, that are denied a territory. These birds usually join roaming flocks.

Visual displays

Magpies are not known for giving conspicuous visual displays. Because pairbonds between most magpies are long-term, there is little need for elaborate courtship routines. Males, however, courtship feed females and females solicit copulations by calling and displaying to their mates. To solicit sexual attention, a female crouches on a branch, spreads her wings, and rapidly fans her tail side-to-side.

Bill down
From Brown, , E.D. & Veltman, C.J. 1987.

Most visual displays are associated with mediating territorial disputes. One such display is the *bill down*, which frequently precedes an attack. The performance involves a bird erecting feathers on the body, head, and neck. The bill is pointed downward with the wings slightly drooped as the bird leans forward.

Wings up
From Brown, , E.D. & Veltman, C.J. 1987.

During disputes, neighboring magpies may fly (two to three meters off the ground) along a shared boundary, for some 50m or so, before fluttering to the ground. Another way of getting attention is tilting the wings side-to-side while flying (five to seven wing beats per tilt).

To convey aggressive intentions while on the ground, a magpie lifts its wing into a "V," an action that exposes its white wing patches. In the event that opponents make contact, they peck, wrestle, and dive-bomb each other. Most disputes, however, are resolved by singing, at which magpies excel.

Play

Australian Magpies are among a handful of birds that are known to vigorously play. Both juveniles and adults pull at each other's feathers and, occasionally, roll on the ground, grab, and carry items while hanging from branches. Elements of sexual behavior may also be exhibited while playing.

Initiating a bout of play

Vocalizations

Both sexes sing. Australian Magpies mimic the sounds heard during social interactions. This is why hand-reared birds readily mimic the sounds of their handlers. In the wild, magpies are prone to learn songs from their parents and neighbors.

Magpies can also invent new song syllables throughout their lives.

Dawn (nocturnal) song: A simple, monotonous song sung at night and early morning by males to advertise their territory.

Warbles: Low pitched and melodious calls of variable duration. Often followed by carols. Individual or group specific; rarely shared between groups. Promotes cohesion.

Carols: Loud and long notes emitted after warble calls. Used in a variety of contexts. Birds in a group often carol together.

Down sweep: A descending whistle sung during chases and battles.

Two-tone: Two descending whistles given in tandem. Occurs during chases and battles. Down sweeps and two-tone calls do not occur together in the same population.

Rally Call: A high pitched alarm call used while mobbing a predator or during an aerial chase.

Yodel: A harmonically complex call that indicates the sender is interested in what is going on.

Habitat

Common in urban landscape and widely distributed in open terrain wherever there are trees for nesting. A Magpie's open cup nest is made of sticks and placed on the outer branches of the upper canopy of trees.

range

Mistletoebird

(length 10-11 cm)

male

DS

Dicaeum hirundinaceum

Flowerpeckers are common throughout the southwest Pacific, but in Australia the only representative is the Mistletoebird. Although Mistletoebirds consume nectar, pollen, and a variety of small invertebrates, their uniqueness relates to their predilection to eat mistletoe berries. In fact, evolution has modified this species' physiology and behavior to accommodate its dependence on mistletoe as a food source.

For example, the muscular gizzard of Mistletoebirds is not used to crush the seeds but positioned in such a manner as to allow seeds to quickly pass through the digestive tract unharmed (typically within 25 minutes). Sustenance is obtained by squeezing the fruit to release the nutritious and very sticky pulp of the mistletoe berries.

mR.

When defecating, Mistletoebirds carefully deposit the sticky seeds of the mistletoe plant on a branch where they can sprout.

To cultivate more seed-producing plants, Mistletoebirds do not randomly defecate seeds on the ground. When excreting a seed, they place their rear on top of a branch. In so doing, the seed's sticky coating adheres to the top of the branch where it is free to germinate. The

female

WJD

bird then steps carefully to the side to avoid knocking the seed off the branch.

Mistletoebirds build a pear-shaped nest that is hung from a twig and woven from plant down and spider silk. The nest's entrance is a collapsible slit that closes after a bird enters and leaves the nest.

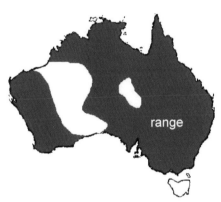

range

Habitat
Found in a variety of habitats wherever mistletoe grows. Outside the breeding season, Mistletoebirds roam over large areas looking for food.

Notes

Adaptive attributes
and useful facts

"Considering my alternative, I'd love to
try an interspecies relationship."

Plumage colors and patterns

male   female

An unusual parrot

Eclectus Parrots are unusual. Not only are females more brightly colored than males, but chicks molt directly into their adult colors. Eclectus Parrots also breed cooperatively—a rare trait among parrots. And lastly, females can determine the sex of their offspring. One female, for example, produced a string of 12 male eggs before switching to lay 13 female eggs. Apparently, a female chooses to produce sons when there is a need for helpers and switches to daughters when she is not having problems raising young.

These bizarre traits are linked to difficulties in breeding in the tropical rainforests of northern Australia and New Guinea. For starters, there is stiff competition among females for useable tree-hollows for nesting. Eclectus Parrots search for hollows located in the tallest tree in the forest because such hollows are less vulnerable to predators (e.g., pythons). A good hollow must also remain dry during the wet season, or nestlings will drown during heavy downpours. Once in possession of a hollow, a female will remain nearby or in the nest for eight months of the year in order to prevent other parrots from usurping the hollow. Even though a female depends upon males to feed her and her chicks, she remains aggressive toward males trying to enter the nest.

From a male's perspective, females that possess a hollow are those that can reproduce. Perhaps this is why multiple males (up to 11) may feed each breeding female—some of these helpers are sexual partners (a male may tend more than one female), but most are offspring from previous clutches. Females do all of the incubation and brooding of chicks; whereas males spend the majority of their time foraging for fruit, nuts, and other vegetable matter. The male's green plumage makes him difficult to spot in the forest canopy.

Brightly colored rail chicks

Bright colors convey information. For adult birds, conspicuous colors attract the opposite sex and warn rivals to stay away. In contrast, the plumage of young birds is usually dull or cryptically colored. Exceptions to this pattern occur when the appearance of chicks evokes a feeding response from their parents. A colorful gaping mouth of a nestling songbird is one example; the red head of a rail chick is another.

The red head of young Dusky Moorhens attracts the attention of adults who might feed them.

After comparing the behavior of 97 rail species, biologists discovered that head ornamentation on young birds is most common in species where adults feed only the youngest chicks—as a young rail becomes less dependent on its parents for food its bright colors fade. In other words, as competition between chicks for an adult's attention wanes the need to visually stand out decreases.

Understandably, competition between young chicks is greatest in large broods, and with regard to rails, large broods can result from joint nesting of females, crèching of young, high rates of extra-pair paternity, and egg dumping where one hen lays its eggs in another hen's nest. All of these possibilities have the same result, namely the occurrence of unrelated chicks being raised in the same brood. In turn, the need to be effective only increases when it comes to begging for food, either by vocalizing or displaying bright colors. Of the two options, the display of a red head and beak seems to work best for Dusky Moorhens.

Tails advertise quality

Male White-winged Fairy-wrens (see photo on page 135) obtain their striking nuptial plumage in their fourth year. Until then, young males resemble females except for sexually mature males that exhibit a bright blue tail. Why do some young males have blue tails? To learn more, researchers measured various attributes of tails from different males. They discovered that the blue-tailed males were large and in better physical condition than brown-tailed males. Also, blue tails not only reflect light in the UV-blue range of the spectrum, but on average are shorter than brown tails.

So it would seem that the length and color of a male's tail can convey useful information, namely whether a male is sexually mature (by color) and, to a lesser degree, information on a male's physical well-being. Since conducting aerial maneuvers are more difficult with a short tail, a male could potentially show off his flying skills more effectively if he possesses a blue tail (also see discussion below).

Melanie Rathburn

Blue and brown tails of two fairy-wren males. The two images were digitally modified to be displayed side-by-side; the actual size of the tails shown here are not to scale.

Some like it short—tails, that is

Charles Darwin believed that elaborate traits such as long tails evolved to help males attract females; he also argued that sexual selection can reduce the size of a trait. This is apparently true for Golden-headed Cisticola. In this polygynous reed warbler, the length of the male's tail after molting (timed to occur just before breeding) is far shorter than the female's tail (the sexes are similar in other respects).

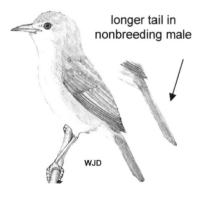

Male Golden-headed Cisticola

longer tail in nonbreeding male

WJD

99

Apparently, a short tail is aerodynamically efficient during prolonged territorial skirmishes. As a result, short-tailed males have an advantage over long-tailed males. With energy to spare, a short-tailed male can devote considerable time courting females, leading to multiple females nesting on his territory. A sexy appearance plus energetic performances could explain why short-tailed males, on average, sire more young than males with long tails.

Coverable badges

A bird's appearance often advertises its social ranking. In this regard, the size and color of specific areas of the body serve as badges that announce the wearer's status. To work, a badge should be a reliable indicator of a wearer's age, health, or fighting ability.

Lewin's Honeyeater

Sulphur-crested Cockatoo

Brush-turkey

Though the physiological costs of growing a plumage badge is low, the social cost of advertising high status can be substantial. Why? Because flashy males are most likely to be challenged by other males. This is one reason why inexperienced young males exhibit duller colors than dominant males.

One evolutionary solution to the problem of attracting unwanted attention is to cover your badge when you are at a disadvantage or do not want to pick a fight. A male Brush-turkey (bottom left), for example, can expand and retract its wattle depending upon prevailing circumstances. The erectable facial tufts of various honeyeaters (above) and head crests in cockatoos (top left), are two additional examples.

The case of the missing crests
In the mating game, convincing the opposite sex to cooperate is often difficult. Among birds, it is usually the male's task to persuade a female that he is worthy of her attention. This is accomplished by dancing, singing, and exhibiting flashy feathers. So far so good, but why do specific traits attract a female's attention in the first place?

For various reasons a bird's central nervous system is biased to detect certain colors or sounds. Assume, for example, that the visual system of a species is sensitive to the color blue because the ability to see blue helps birds locate food. Because of this sensory bias, it is possible that females looking for a mate will prefer blue males (e.g., Satin Bowerbirds). If this happens, there would now exist strong selection for males to exploit this sexual preference by evolving blue feathers. If this intriguing idea is valid, it could explain why traits that seem to have no special survival value can nonetheless affect the behavior of the opposite sex. Take for instance what happens when you modify the appearance of male grass-finches.

A Long-tailed Finch (left) and a Zebra Finch (right) are fitted with artificial crests.

No species of Australian grass-finch exhibits a crest on the forehead or nape. Yet when researchers attached white, red, and green feathered crests on heads of Zebra Finches and Long-tailed Finches, the effect on the behavior of females was amazing. In particular, females of both species acquired a strong and Lasting preference for white-crested males. The effect on males was mixed. Male Long-tailed Finches also preferred females adorned

"At this rate, we'll never get fed while Lucille wears the hat with the purple feather in it!"

with a white crest, whereas male Zebra Finches rejected crested-females regardless of the crest color. Possibly, a sensory-bias for white is latent in both species. Could this be why females of both species pad their nests with white feathers?

Concealed identity

Male and female Long-tailed Finches look alike. At first this may seem peculiar given that males benefit by displaying bright colors and flashy adornments such as tall crests. In some circumstances, however, it is better to conceal any trait that reveals social status, including one's sexual identity. Is this why juvenile males of many species look like females? Such "deception" can reduce the level of harassment by adult males who otherwise would perceive young males as sexual competitors.

Consider the following facts: Long-tailed Finches interact frequently with one another within their respective colonies (consisting of 10 to 20 pairs) and within foraging flocks. Also, males and females establish long-term bonds that lessen the need for males to persuade females to mate with them. Still, dominant males approach and display to newcomers in a group. Such males are known to perform upright postures, sing, and even attempt to copulate with new arrivals. Are the resident males probing for information regarding the sex of a newcomer since outward appearances are ambiguous? On the other hand, if you are a new bird in the area, it is not necessarily a good strategy to arouse the sexual jealousy of a more dominant male.

Zebra Finches

Although adult Zebra Finches are sexually dimorphic in appearance, all adults have a red bill, whereas the beak of juveniles is black. In fact, if the bill of a young finch is painted red, its parents will stop feeding it, which indicates that bill color is used to identify the age of a finch. (Note: Older fledglings with red beaks resemble adult females.)

Juvenile (right) begs to be fed.

WJD

Male Zebra Finch

When feeding nestlings, wild Zebra Finches selectively feed only those chicks that exhibit the appropriate species-specific gape marking. This innate selectivity, however, has been lost in some domesticated Zebra Finches. As a consequence, birds bred in captivity will readily adopt chicks of other species.

Symmetry

When wooing a female, the quality of a male's singing performance can be very persuasive. When all the males courting a female are good singers, however, she must make her evaluation based on traits other than a male's song. One possibility is to closely examine a male's appearance, particularly symmetry of his plumage such as the arrangement of the feathers on a male's throat and chest. To produce such an intricate pattern of finely spaced barring, feather growth during molt must be nearly flawless. Any minor disruptions while a bird is molting can result in irregularities in the final pattern.

Though few males are perfect, some are more symmetrical in their appearance than others. During their assessment, females pay little attention to the number of bars or the width of spacing between the bars on the throat. It is more important that the barring pattern flows smoothly from left to right. Breaks in the pattern are decidedly unimpressive to the opposite sex. When given a choice between several males, a female Zebra Finch prefers to mate with the male that is the most bilaterally symmetrical.

Why does appearance matter? Apparently, deviation from perfect symmetry indicates that there were problems during early development. This is important because resistance to environmental assaults early in life is a measure of a male's genetic fitness.

Black and white birds

Pied Butcherbird   Pied Currawong   Pied Honeyeater

Possible reasons for exhibiting black and white

- Black and white birds commonly live in open habitats.
- Black and white birds defend territories year-round using visual displays and vocalizations.
- In a brightly lit environment, the contrast of black on white is effective in emphasizing form and movement.
- White has the highest reflectance value of any color, making it exceptionally good for advertising movement over long distances.

When viewed from afar, the outline of white objects becomes blurred. Perhaps this is why few small birds are entirely white. Black, on the other hand, holds the shape of a bird against most visual backgrounds. Hence, an appropriate mix of black and white feathers highlights subtle movements during courtship displays and when birds are in flight.

Is it a coincidence that those species that advertise their presence by perching at elevated locations for long periods of time are usually black (e.g., crows, Pied Currawongs, Pied Honeyeaters, Spangled Drongos, Australian Magpies, and Magpie-larks)?

In an influential book, Amotz Zahavi recalls an experience of a friend's daughter being scolded for not wearing black leotards during dance practice. Black allows an instructor to monitor the movement of a dancer's legs. In a live performance, however, dancers wear white leotards to limit the ability of the audience to pick out mistakes.

To humans, black and white feathers are perceived as two colors. This is not necessarily the case for birds since many black and white feathers reflect a fair amount of ultraviolet light that birds can see. Hence, birds may not perceive all black or white feathers as being the same color. In fact, females of some species prefer to mate with males that exhibit not only large black status badges but feathers that also reflect large amounts of UV light.

Vocalizations and sounds

Booming cassowaries
Pigeons can detect ultra-low sounds such as those produced by crashing waves and thunderstorms. Because low frequency sounds have long wavelengths, they can travel long distances with relatively little loss of energy. Likewise, low pitched sounds can penetrate dense vegetation.

Actual production of ultra-low sounds by birds is known for only a handful of species including the Eurasian Capercaillie and the Dwarf and Southern Cassowaries. The booming calls of both cassowaries contain significant energy in the 23 Hz and 32 Hz range. Presumably, cassowaries emit ultra-low sounds to communicate over long distances and

Southern Cassowary (female)

also through the thick understory of their tropical forest habitat.

Some researchers speculate that the large keratinous casque of cassowaries functions as a listening device. Accordingly, the casque in living birds is spongy and resilient, and within its deeper regions there is an ample deposit of "darkly pigmented sludge." The semi-solid interior of a casque could be used to detect the low-frequency vibrations generated by booming calls. Alternatively, a bird's casque could be used to attract the opposite sex, to wield as a weapon when fighting, to scrape leaf litter, or to protect the head as a bird crashes through vegetation.

All of these functions, however, can be easily refuted by watching how cassowaries behave in the wild.

Catching wolf spiders

Most Restless Flycatchers live in open habitats. They differ from other so-called monarch flycatchers in several important ways. The sexes look alike and exhibit an unusual mode of foraging for insects, worms, and spiders. In their quest for food they routinely hover a few feet above the ground and occasionally along the outer edge of trees when inspecting foliage for invertebrates. While hovering,

they often give a harsh grinding sound that is so low in pitch that it seems an unlikely sound originating from such a small bird.

It is plausible that the scissor-grinding call helps Restless Flycatchers catch prey such as wolf spiders that live in subterranean burrows or in the hollows of trees. How? By the sound resonating within the burrow that, in turn, causes the spider to approach the burrow's entrance. When the spider becomes visible to the flycatcher, it also becomes vulnerable to attack.

Restless Flycatchers (above and left) that hover and emit a scissor-grinding call are most likely searching for prey. In this instance, the bird on the left is foraging for wolf spiders.

Singing together

When two or more birds "intentionally" sing together, they are said to be duetting. The vocal duets of birds are often coordinated with visual displays, and not all duets require a bird to sing. In fact, some birds duet using nonvocal sounds as illustrated by the drumming sessions of woodpeckers and the bill-clattering of storks and albatrosses.

An intriguing feature of some avian duets is the precise orchestration of the voices. Often participants sing at precisely timed intervals during the performance. So precise are some duets that it is difficult to discern if one or two birds are singing. Most duets, however, are more free flowing; in many instances, the songs of a male and a female simply overlap one another.

Duets may last for many seconds with each bird alternately jumping in and out of the performance (e.g., Magpie-larks). Or one bird may simply answer the other as exemplified by the whip-cracking calls of male Eastern Whipbirds.

The fact that 80 percent of duetting species reside in the tropics suggests that there is something about living near the equator that causes females to sing. For starters, tropical birds are sedentary with both sexes defending territories year-round. Because singing is a cheap way to defend the home turf, it makes sense that females sing as well as males. Furthermore, for the majority of species that duet, males and females look remarkably alike—a fact that helps a pair jointly defend a territory.

Okay, so being able to sing is useful in defending a territory, but why do males and females sing together? There are several plausible reasons. Coordinated singing can strengthen a couple's pairbond. Pairs may duet to advertise to competitors that their territory is occupied or to ward off individuals looking for a partner. In the latter context, duetting could represent a form of mate guarding. Below are a few examples of Australian birds that duet.

Magpie-larks

Throughout the year, Magpie-larks sing and visually display together, often on the same perch. Are duets explicitly performed to strengthen the pairbond or ward off sexual rivals? Apparently not, because neither sex goes out of its way to answer their partner's song during two critical periods: when the female is fertile and when a pair is separated. These facts suggest that the duets of Magpie-larks are performed primarily to defend real estate. Singing and displaying from an exposed perch, either alone or in a duet, maximizes the gain from one's effort.

Budgerigars

The popularity of budgies as pets stems in part from their ability to vocally mimic their caregivers. As pet owners are aware, some budgies are silent while other are talkative. Such disparity is related both to the sex and up-bringing of the birds. Researchers have revealed

male   female

Magpie-larks often accompany their visual displays with vocal duets.

that males reliably imitate contact calls of their mates. Moreover, those males that invest the most in head bobbing and feeding the female are

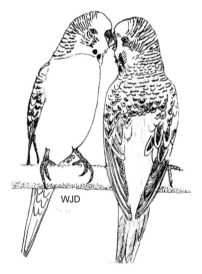

While courting, male budgerigars learn to mimic a female's contact call.

invariably the most accurate and vociferous imitators. It so happens that imitation of human speech is derived from the natural tendency of males to imitate the contact calls of females. If having a talking budgie is your goal, then select a hand reared male that has imprinted on humans. It is also recommended that you regularly interact with your pet.

Eastern Whipbird

Males sing simple but impressive songs comprised of a long tonal whistle followed by a loud crack that resembles the snap of a cattleman's whip. If paired, the male's song is usually answered by the female with multiple *chow chow* notes. On rare occasions, a single bird, presumably the male, sings both parts of the duet. The reason he does this is unknown.

Why is the male's song composed of multiple components? Presumably the whip-crack unambiguously advertises the male's vigor and location. (If he sings outside his territory, his mate typically does not respond.) But why whistle at the beginning of the song? Is it necessary

Eastern Whipbird

to whistle to produce a respectable crack? Possibly. There is another clue, however. Males can and often do vary the pitch of the introductory whistle, especially when multiple males are calling. This makes the introductory notes useful when counter-singing with other males.

Common Koel

Though normally a rainforest species, the Common Koel, a parasitic cuckoo, has become a common sight along the eastern seaboard of Queensland, especially where large numbers of fruiting trees, particularly figs, have been planted. Staying high in the canopy, koels are rather difficult to see among dense foliage, but their incessant *coo-ee, coo-ee, coo-ee* is conspicuous. Koels often sing for hours throughout the day and night.

When koels first arrive in spring from Papua New Guinea, the average number of calls given per hour is about 125; in late summer, the rate decreases to less than 20 per hour. Are males singing to defend territories? Not likely, because individual males do not remain in a particular location for more than a few weeks. Presumably, once all available nests are parasitized, koels move to a different location to search for more nests.

It would seem that singing is used to establish a male dominance hierarchy among the local males as they compete for calling posts near the nests of potential host species. Assuming that dominant males

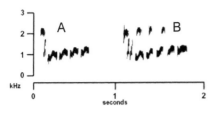

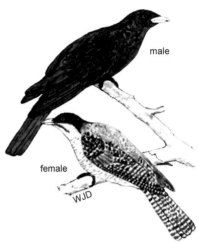

(A) Call of a male koel

(B) Duet shows the high pitched call of a female overlapped by the lower pitched call of a male.

attract more females, Common Koels should be considered polygynous.

The above description of the koel's social life seems to roughly fit with observations made in the field. Recent research, however, has found that Common Koels also duet, a total surprise given that most known duetting species are socially monogamous.

Why should females sing and duet with males? By singing, a female can advertise to other females that the local area is being mined for nests; also, by duetting, a male and female can establish a pairbond, albeit brief. If correct, koels are, in effect, practicing serial monogamy.

Nonvocal sounds

Sounds produced by wings, tails, beaks, and feet can, in some circumstances, convey meaningful information. Classic examples are the sounds created by the uniquely shaped wing and tail feathers of doves and snipe.

In northern Queensland and throughout Papua New Guinea, Palm Cockatoos beat the trunks of hollow trees with a variety of items: a branch (cut for the purpose); a stone; or a large seed capsule. Drumming by cockatoos is heard most often early in the breed

Palm Cockatoo drumming

ing season, presumably to claim title to a territory. Also, they may bang on potential nesting trees to evict current occupants (e.g., other birds, mammals, or even a predator) from the cavity.

Wings
Flapping wings inevitably produce noise with no signaling function There are possible exceptions such as the flip-flop sound of Eastern Spinebills, the rustling of a displaying Magnificent Riflebird, and the whirring flights of Chowchillas. Rock Warblers also produce a spinebill like whirring with their wings during territorial disputes.

The wings of a number of pigeons (e.g., Bar-shouldered Doves and Crested Pigeons) produce sound when flying, during takeoff, and during aerial displays. In doves, such sounds are produced by air rushing over oddly shaped primaries. In addition, a few species, such as Feral Pigeons and Wonga Pigeons, slap their wings together over their back, producing a loud clapping.

Tails
Each of the three species of snipe in Australia makes precipitous dives from great heights during aerial courtship displays performed on their breeding grounds. On the steep descent, air rushing over the outer tail feathers creates a tremulous, bleating sound lasting several seconds.

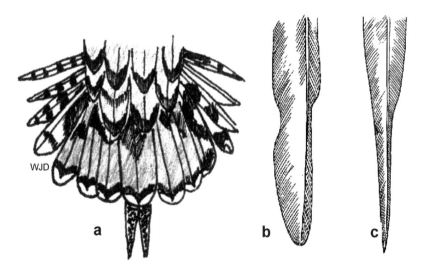

The thin, outer tail feathers of the (**a**) Latham's Snipe vibrate as air rushes over them during the male's aerial display. Oddly shaped wing feathers of the (**b**) Common Bronzewing and the (**c**) Crested Pigeon generate whistling sounds during flight.

Bills

Sounds are also created by bill-snapping, bill-rattling, and bill-fencing. When confronted with a predator or a rival, many birds issue loud cracks by quickly closing the bill. In a less adversarial context, bill-rattling is used in courtship dances, nest exchange ceremonies and appeasement displays of a number of species (e.g., storks, cranes, gannets, pelicans, and albatrosses). In some instances, bill-clapping and vocalizations are used together (e.g., Marbled Frogmouth).

Bill-fencing is practiced by grebes, kingfishers, herons, pigeons, crows, penguins, and some songbirds such as Figbirds. It is unknown if the sounds produced during bill-fencing have any significance or if they are purely incidental.

Feet

Rarely are feet the source of nonvocal sounds in birds. Possible exceptions of Australian species include drumming by Palm Cockatoos and bush-shaking by Albert's Lyrebirds. In both cases, the feet are used to grasp an object (a tool?) to produce the sound.

"Oooh, I can just imagine the size of his territory."

Breeding and nesting

Nest associations

Non-aggressive birds nesting near or within the nest of a bird of prey

Diamond Firetails and Yellow-rumped Thornbills sometimes nest within stick nests of Little and Wedge-tailed Eagles; likewise Zebra Finches occasionally nest in the nests of Spotted Harriers.

Single pairs of unrelated species

Willie Wagtails, Restless Flycatchers and White-winged Trillers nest near Magpie-larks; Brown Thornbills nest near White-browed Scrubwrens (New South Wales); and figbirds, drongos and Helmeted Friarbirds (Magnetic Island) nest within the same tree.

Other nest associations include Trumpet Manucode and Black Butcherbird (Cape York), Regent Honeyeaters and Noisy Friarbirds, Striped Honeyeater and Grey Butcherbirds, Leaden Flycatcher and friarbirds, and Leaden Flycatcher and Olive-backed Orioles.

Birds associating with social insects

Double-barred, Red-browed, and Spice Finches nest near paper wasps, as do several species of gerygone. Birds living in termite mounds include species of kingfishers, parrots, and pardalotes.

WJD
Buff-breasted Paradise
Kingfishers (above) and
Golden-shouldered Parrots
(right) nest in termite mounds.

WJD

Egg tossing and mate sharing in moorhens

Multiple male and female Dusky Moorhens may share a nest, which can lead to some bizarre behaviors. One female, for example, after laying an egg, was observed shoving an egg out of the nest and then trying to retrieve it, unsuccessfully, from the bottom of the lake.

In another instance, a clutch of eggs was laid and then abandoned, for unknown reasons, by all members of the group except for one male. When the females started laying eggs in a new nest, the male that remained with the first clutch of eggs repeatedly visited and knocked eggs out of the new nest. Why trash the second nest? Since the offending male had not fertilized any of the newly laid eggs, he was probably protecting his only investment in the first nest.

Why do moorhens share a nest? If there are more adult males than females in a local population, a male may be better off tolerating other males than risk not mating at all. Females may band together because it is difficult to defend a territory; related females would have an even stronger incentive to cooperate in raising young.

Producing offspring may not be the only reason to cooperate, given that access to a good territory provides its own perks. Food within a territory is one advantage for joining a group; protection from predators is another.

Mounds as incubators

Most birds use body heat to incubate their eggs. Megapodes (e.g., Australian Brush-turkey, Malleefowl and Orange-footed Scrubfowl), in contrast, bury their eggs in mounds of vegetation where they are warmed by fermenting plant material.

Cross-section of a Malleefowl mound (0.5 to 1m high). Decomposing vegetation is typically located within an area below and slightly above ground level (outlined with dashes).

Only males build mounds. After a female lays the eggs, the male controls the internal temperature of the mound by adding and removing vegetation. Mound building probably originated from ground-nesting ancestors who sat on their eggs. During periods when the birds left the nest, vegetation may have been thrown over the eggs to hide them from

Brush-turkey on mound   Malleefowl

predators. Presumably, over generations, the habit of covering the eggs evolved into mound building and eventually led to eggs being laid directly into the mound to be incubated. The freedom of adults to abandon their eggs produced strong pressure on chicks to not depend on inattentive parents. In fact, newly hatched megapode chicks are highly precocial. Immediately after digging its way out of the mound, a Brush-turkey chick can feed itself, thermoregulate, run and even fly.

Building a nest out of mud

Have you ever thrown a vase on a potter's wheel? It is not as easy as it looks. Now imagine the time, patience, and skill required to build a reasonably symmetrical nest made of mud.

Three Australian species, the White-winged Chough, the Apostlebird, and the Magpie-lark, have perfected the technique of building nests with wet mud. All three species balance the nest on a horizontal branch.

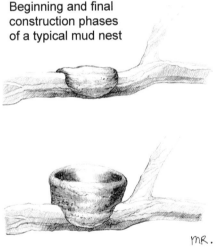

Beginning and final construction phases of a typical mud nest

MR.

Little by little, mud is deposited in layers to form the nest's platform (*top right*) and walls (*bottom right*). In hundreds of trips to the nest, beaks full of mud mixed with bits of vegetation are molded into the walls. On each trip, the bird taps and vibrates the wet mud with its bill to expel trapped air. Next, to assure that the nest is properly shaped, the builder sits in it and vigorously presses its body against the sides. Before eggs are laid, the bowl is lined with soft material that provides cushioning and insulation.

Because White-winged Choughs and Apostlebirds are closely related species, it is possible that they inherited the practice of mud-building from a common ancestor. Neither species, however, is related to the Magpie-lark in which mud-building evolved independently. Still, there could be a shared reason why each species builds mud nests. At present we can only guess what it might be.

Three mud nesters

White-winged Chough (above)

Magpie-lark (above) and Apostlebird (below)

Magpie behavior

Some urban dwelling Australian Magpies dive-bomb humans. Cyclists riding past the nest tree, for example, often evoke such defensive behavior but, fortunately, not from the majority of magpies. Why do some birds swoop humans while others do not?

Swooping magpie

Magpies, almost always males, mob humans during the breeding season, suggesting that the birds are defending the nest and young. Furthermore, the majority of swoops occur in public locations where the birds frequently encounter unfamiliar humans, as opposed to the familiar humans in someone's backyard. Typically, wild magpies living in the bush that have little contact with people do not mob but instead retreat when a human approaches the nest. This response suggests that magpies do not instinctively dislike humans, but learn to direct their attention toward a specific class of people.

Familiarity, however, can both reduce and increase swooping by magpies and other species such as butcherbirds. Presumably, young magpies can learn that humans are a threat by watching their parents; whereas other birds may have personally experienced harassment from humans, leading to mistrust. In both circumstances, magpies learned to identify humans as nest predators.

Recently fledged magpies often wait on the ground for their parents to feed them, and if a nest is near a schoolyard, children sometimes pick up the young birds believing that they have been abandoned. From the magpie parents' perspective, such actions are perceived as a direct assault on their young. To defend their young, the parents mob the children by swooping past the perceived predator's head and neck. Such behavior can have serious consequences if contact is made because magpies possess a large, sharp bill that can easily cut the face, back of the neck, and even injure an eye. Hence, in most areas, local authorities are known to capture and relocate troublesome males.

Caring stepfathers

When a breeding male dies or is forcibly removed from his territory by humans, what becomes of his offspring? Would his mate be able to feed the chicks on her own? Would the neighboring magpies take over the pair's territory? Or, if a female finds a new mate, would the new male attempt to kill the female's chicks in order to induce her to renest?

Researchers discovered that widowed females usually acquire a new male within a day (sometimes within hours) after the disappearance of their mate. Surprisingly, replacement males are usually not a neighbor, an already paired male, or a male that owns his own territory. The newcomer simply appears from the ranks of floaters omnipresent in the neighborhood, possibly a brother of the original male. More research is required to identify the social dynamics involved.

Females treat their new mates the same as they do their previous partners. Furthermore, most replacement males quickly start singing and patrolling territorial boundaries. More impressively, replacement males immediately begin bringing food to their adopted chicks! In fact, they provide twice as much food as the biological father did.

Instead of killing unrelated chicks, a replacement male may fare better if he adopts a long-term perspective. Since magpie pairs jointly defend the breeding grounds for years, breeding vacancies within a local area are rare. In this context, foster care may be a sensible approach for a young and unproven male. Furthermore, an over-the-top exhibition of chick-provisioning abilities may help persuade the resident female that the newcomer is suitable to keep around to help raise her next brood.

The power of the crèche

Raising a brood of chicks can be demanding work. Usually chicks are left alone in the nest while their parents look for food. Alternatively, the young join together to form a crèche, one large brood that is looked after by one or two adults. The joining of broods has been reported in a number of species including terns, pelicans, flamingos, some parrots, penguins, ostriches, swans, and ducks.

To date, there is no evidence that families that form crèches are cooperating with one another or that relatives are helping each other out. To the contrary, abandoning one's own chicks can be a "cheater" strategy designed to exploit the parental behavior of others. Is this really what is happening?

Scarce food and avoiding predators

If, for example, food is difficult to find, it makes sense to leave flightless young in a safe place while the parents depart to search for food. Also, when predators are prevalent, it makes sense to combine several broods since there is relative safety in numbers.

Rather than leave newly fledged chicks to fend for themselves, Galahs escort their offspring to communal roosts occupied by other juveniles. At these roosts at least one parent stays near the crèche.

A similar approach is used by colonial nesting pelicans (see figure below) and penguins. Pelicans leave their chicks in nurseries located near the breeding colony.

When multiple families of ducks, geese, and swans converge on the same feeding sites, different broods may fuse with little or no aggression between adults. Upon the parents' return, the families reunite. In some circumstances, however, territorial conflict may force a female to abandon her offspring, not out of choice, but because her chicks are unable to follow her retreat. Caring for the chicks is now the victor's responsibility. Such an outcome may still be a win-win situation since joining a crèche often increases a chick's chances of survival; in fact, chicks raised in large crèches often grow faster and fledge sooner than those raised in single broods. Such perks give crèche-raised birds a post-fledging advantage over single-brood chicks.

Other considerations

Under hardship some hens will abandon their first brood, giving them the opportunity to renest. This is true if the cost of raising young is independent of brood size, a hen's brood is small, and if the care of a hen's chicks is assumed by another hen.

Under the age of ten days, most hens are unable to recognize their own chicks, especially if the foreign chicks are similar in size to the hen's own offspring. There are two rules of thumb to consider before becoming a foster parent: first, if a chick is the appropriate size, then defend it against predators because it could be your own; second, if the cost associated with adopting unrelated chicks is low and predators are about, then welcome new arrivals because there is safety in numbers.

Featherless and flightless, young pelicans huddle in pods when their parents are away feeding. Such behavior conserves body heat and protects very young birds from predators.

SLN

- Chicks, three to five per nest, fledge over a period of two weeks. On a chick's maiden flight, the parents fly alongside the young bird.

- Parents escort fledglings to communal sites, often several kilometers from the nest, where other juveniles are roosting.

- Locations of crèche sites are influenced by tradition.

Young Galahs wait
for their parents to return.*

- As many as 50+ young may congregate in a crèche where young birds practice flying, respond to each other's alarm calls and wait for their parents to return with food.

- When returning with food, parents call in flight to encourage their offspring to assemble in a particular tree.

- Juveniles remain in a crèche overnight while the parents return to their nest hollow to roost.

- Most Galah families change crèche areas at least once before the young are completely independent.

- Nursery areas serve as an assembly point until all of the young in the nest have fledged, at which time the whole family moves to a site closer to the current food source.

- At approximately five weeks after fledging, the young leave their parents to join roaming juvenile flocks.

* Drawn by Belinda Brooker and used by permission from Ian Rowley

- At low population densities, nesting pairs defend territories and cygnets are attended by both parents.

- In nesting colonies, broods of four or more pairs may amalgamate and be attended by one pair of adults.

- First time breeders may pair only temporarily with either sex partner deserting the brood; this may allow for up to four clutches per year for some females, with nearby broods joining crèches.

- Young hatch asynchronously, resulting in mixed broods with the age between chicks spanning several days.

- In captivity, cygnets from several broods intermingle and only separate when called away by parents.

- Foreign cygnets attempting to join a brood containing larger chicks are attacked by adults (parents) tending the brood.

- On lakes, early broods are escorted to remote areas where they are raised by both parents. Cynets that hatch late in the season join crèches containing chicks from two to four broods.

- Chicks move from one crèche to another with some groups containing young from up to 30 different broods.

- Crèches are most likely to form when cygnets are fed within a small area where food is concentrated. Under very harsh circumstances (i.e., lake dries up) parents may abandon their chicks.

Black Swans nest as single pairs and in colonies. Cygnets of colonial nesting swans often intermingle when adults lead their broods to communal feeding sites. Once together, crèches of fifty or more swan chicks may form. Each crèche is attended by a pair of adults.

Symbolic use of nesting material

Manipulating nest material during court-ship displays has been observed in a number of species including finches, boobies, and grebes. Such behavior is called symbolic nest-building because the items handled are not used in construc-tion of the nest. Courtship in some Australian grass-finches (e.g., Red-browed, Red-eared Firetail, Star, and Crimson Finches)

Courtship of Red-eared Firetail Finch

involves the male singing and performing displays in the vicinity of a female while he holds a long blade of grass in his beak.

When a male has attracted the attention of a female, he leads her to a number of potential nest sites. A hundred sites, however, might be vis-ited before one is chosen, usually in a bush or a small tree. If she finds one of the sites acceptable, she will consummate the union and start to build a nest. Only then does the male proceed to search for material to give to the female for nest building.

Cooperative breeding

In a small minority of species (close to four percent worldwide) multi-ple adults share parental duties. Interestingly, a disproportionately large number of Australian birds are cooperative breeders—approxi-mately 22 percent of the endemic birds. The tendency to breed coopera-tively is more prevalent in some bird families than in others, and it turns out many of these families occur in Australia.

A most common cooperative arrangement is when a single pair is assisted by non-breeding individuals of the same species. The reasons why some individuals forgo their own reproduction to help others have intrigued biologists. Providing assistance could be the best option avail-able to an experienced bird that has not acquired its own territory. In other words, if all available breeding habitat is occupied in a given area, some individuals will be prevented from breeding. Yet, they could do

Examples of cooperatively breeding birds

Australian Magpie	Grey Butcherbird
Fairy-wrens	Apostlebird
White-winged Chough	White-browed Scrubwren
Noisy and Bell Miners	Eclectus Parrot
Laughing Kookaburra	Rainbow Bee-eater

some good by remaining at home to help their mother raise younger siblings. This is a common scenario for many young males (less common for females).

Assisting in raising a relative's offspring is one way of promoting the family line. Yet, in the White-browed Scrubwren, male helpers are usually not related to the breeding female. In this instance, helping has become a sexually selected trait, not unlike bright plumage and elaborate courtship displays. If females are prone to mate with males that demonstrate that they are good providers, a male can increase his chances of mating with a female at some future time if he impresses her by helping to raise her current batch of chicks.

As mentioned previously, becoming a helper could enhance a young bird's status within the group, especially if assisting comes at a cost to the helper. In White-winged Choughs, the cost of helping may be too high for some young birds. This is why they decide, on occasions, to eat the food they bring to the nest, especially when the resident breeders or other helpers are not watching. Assuming a helper does not get caught cheating, refusing to feed the breeder's offspring reduces a young bird's

(Continued on page 126)

Several unrelated males might help feed the offspring of a female White-browed Scrubwren. They do so to gain an opportunity to mate with her at some future date.

Adult Grey Butcherbirds take turns feeding nestlings. At times, a helper brings food to the nest without feeding the chicks, especially when it arrives alone.

Explanations for helping in birds

- Payment of rent or mutualism: A helper is permitted to stay on the territory and receive the benefits of group living as long as it provides help to the territory owners.

- Access to potential mates: Helpers are trying to gain direct access to reproductive opportunities either within or outside of the social group.

- Gain prestige: Helping might improve an individual's social standing which, in turn, may improve mating opportunities.

- Improvement of local conditions: Helpers are accepted to increase group size; a large group can expand or improve its territory to increase its reproductive output.

- Establish strategic alliances or coalitions: Helping promotes the formation of alliances between individuals that enhance the chances of acquiring future breeding opportunities.

- Gain experience: Helping gives birds the chance to practice critical parenting skills.

- No cost to help: Helping may provide no benefits. It is simply a behavioral response to the presence of begging young.

For some young White-winged Choughs, food can be difficult to find. The hungriest helpers attempt to cheat and eat the food themselves, but not before they pretend to feed the nestlings. False-feedings occur most frequently when other choughs are not watching.

costs of finding its own food while upholding, albeit deceptively, its social prestige within the group. Cheating has also been observed in families of Grey Butcherbirds and Bell Miners. In miners, as many as 20 helpers may provision the young of one female at a rate that far exceeds the chicks' needs. Why so many? Part of the explanation is linked to the fact that there are significantly more males within a colony than females. Hence, most males do not have an opportunity to mate.

In general, females are usually much choosier than males when selecting a sexual partner. Hence, one would expect to find more bachelor males willing to become helpers than females—this is the case in the vast majority of cooperative breeders.

Why do pardalotes burrow?

Pardalotes have two options for nesting: a pair may dig their own burrow, usually in an earthen bank, or they may choose to nest in a tree-hollow. Where a pardalote nests depends less upon the availability of hollows and more on the composition of the local bird community—in particular, on the presence of other birds that compete for the same food source.

A Striated Pardalote
leaving its burrow.

Pardalotes, like honeyeaters, eat sugary excretions of psyllids, small plant sucking insects (also called lerps). As many as 18 species of honeyeaters, including the Yellow-faced and Singing Honeyeater, sample lerps when available; some honeyeaters, such as Bell Miners, are lerp specialists.

Lerps are a stationary food source zealously defended by territorial honeyeaters, a fact that promotes conflict between honeyeaters and pardalotes. Most honeyeaters readily chase and can even kill pardalotes when protecting the local supply of lerps.

Pardalotes can reduce harassment from larger birds by feeding in flocks, as they do in the nonbreeding season, and by excavating nest burrows low to the ground. In localities where honeyeaters are scarce, pardalotes have the option of nesting in tree-hollows.

The nest of most Striated Pardalotes is placed at
the end of tunnel excavated in an earthen bank.

Response to predators

Brush-turkey

It is an impressive sight to watch a Brush-turkey chick emerge from a mound of leaves and dirt. The chick hatches within the mound and digs its way to the surface, a process that can take up to 48 hours. Once free, the highly precocial chick is en-

Young Brush-turkey leaves its mound.

tirely on its own. It must find its own food and escape from unfamiliar predators.

Researchers presented different types of predators to two-day-old chicks: a live cat, a live dog, a rubber Red-bellied Black Snake, and a silhouette of a Grey Goshawk. The alarm calls of several species of songbirds were also played to the young birds.

When confronted by either a dog or snake, the chicks attempted to run away; when confronted by a cat or a raptor the young birds crouched down and froze. When the models were removed, the birds had different responses. The chicks remained crouched for a longer period after removal of the raptor than removal of the cat. When the snake was removed, the chicks soon stopped running but continued running after the dog was removed. And lastly, when the alarm calls were played, the birds stopped what they were doing and visually

Hatchling Brush-turkeys are superprecocial. When they crawl out of the mound they are totally independent, fully feathered, and able to fly.

128

Galahs (above) escape aerial predators by out-flying them; whereas Red-rumped Parrots (left) mob their adversaries or head for cover.

DS

WJD

scanned the immediate area; when the calls stopped, the chicks resumed their normal activity. (Note: Adult Brush-turkeys do not emit alarm calls). Though Brush-turkey chicks probably do not recognize specific types of predators, they react appropriately to the size, speed of movement and height of items that pose a danger, real or otherwise.

Parrots

How birds respond to a predator can vary dramatically, even among closely related species. Red-rumped Parrots, for example, often mob predators and then head for cover. In contrast, Galahs attempt to outpace and out-maneuver an aerial predator. They generally avoid flying into the trees where aerial maneuverability is hampered and ambush is possible. They rely instead on their acrobatic flight skills and endurance.

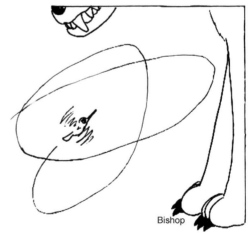

Bishop

"C'mon put-em ... Wassa matter? Chicken?"

129

Mobbing by Noisy Miners

When a predator is spotted in a tree, Noisy Miners invariably mob it. Are they defending a nest, or is there more to the story? Biologists discovered that 80 per-cent of mobbers are nonbreeders that are unrelated to the miners whose nests they help defend.

Mobbing a Pacific Baza

When stuffed models of predators (e.g., a crow, Falcon, and dove) were placed near active miner nests, several trends emerged. Significantly more miners mobbed the model of the falcon than the crow, suggesting that a falcon represents a greater threat to miners than a single crow. Surprisingly, a similar number of birds were observed attacking the dove and the falcon. As expected, more miners mobbed the models early in the breeding season as opposed to later when parents were busy feeding nestlings.

How can these results be explained? Apparently, Noisy Miners are prone to mob a predator when its identity is unknown, the miner is threatened, or when there is no conflict of interest such as feeding young.

A more controversial idea contends that Noisy Miners mob predators as a form of pseudo-reciprocity which could explain how neighboring miners mutually benefit when disposing of the threat. Alternatively, the act of mobbing could convey information to other miners such as "see how fit I am!" In effect, by helping defend the nest of a non-relative, a Noisy Miner can show off to other miners. In this scenario, the benefit of participating in cooperative displays is the elevation of social status and prestige within the neighborhood.

Fairy-wrens

When Superb Fairy-wrens encounter a predator, they may emit alarm calls and/or perform distraction displays. There is also a third, somewhat unusual, response of males. Some males may sing a special "predator song" upon hearing the call of a raven or currawong, even when the predator is merely flying overhead. Such songs are lower in pitch and structurally simple compared to ordinary songs. As a result, they can be heard over longer distances. All males within a group sing similar predator songs that can be distinguished from those in neighboring groups.

When local males simultaneously respond to the call of a predator, females from adjacent territories have an opportunity to compare the songs of multiple singers. For this reason, researchers believe that predator songs are emitted to impress any sexually receptive female that can hear them. Those males that respond vigorously demonstrate stamina and good health—qualities important to females.

Viewed in this context, it makes sense that predator songs are designed to be heard over long distances and should be sung by males whether they are on or off their territories. Presumably, only males in good physical condition can afford to superfluously respond to aerial predators that are potentially a threat.

A male fairy-wren responds to the call of a raven.

Antipredator behavior

Imagine that you have just spotted a Peregrine Falcon flying along a river. Nearby there are ducks swimming, shorebirds feeding along the banks, and a flock of crows roosting in a tree. In response to the falcon, the ducks and crows move and vocalize. As the threat becomes urgent some birds give alarm calls, perform distraction displays, and take evasive action.

Which response the arrival of a predator evokes depends upon an individual's circumstances. Ducks, for example, that are not attending chicks often dive underwater when a falcon swoops overhead; whereas crows take off and attempt to fly higher than the raptor. If the falcon lands on an exposed perch, birds with young may attempt to distract the predator while the crows are likely to mob it. When interpreting anti-predator behavior, it is helpful to look for contextual clues. A short list of anti-predator behaviors of birds is given below.

Vocalizations: Alarm calls are frequently given when a predator is first spotted. The type of call emitted varies with the context and "intended" function of the call. Some species use distinctive alarms for ground and aerial predators. Other species produce dangerous sounding calls to deceive predators (e.g., mimicking the sound of a hissing snake).

Detection: To detect the approach of a predator, some social species place *sentinels* in elevated positions while flock members forage on the ground.

"Hey, that wasn't a predator"

"I know, but I never liked that guy."

Bishop

132

Wonga Pigeons adopt a cryptic posture that makes them difficult to see against a leaf covered ground within a forest.

A Willie Wagtail mobs a Laughing Kookaburra.

Repelling: Mobbing can be contagious in that individuals from various species often join forces to harass a common enemy.

Avoidance: Two common actions to avoid capture are *fleeing* (e.g., retreat to cover, attempt to out-fly the predator, etc.) and *freezing* (response of young birds to their parent's alarm call).

Distraction displays: "Rodent-run" and "broken-wing" displays are performed to distract the attention of a predator away from offspring.

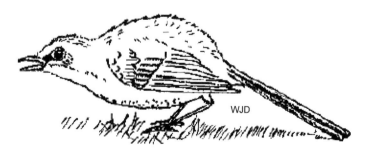

Rodent-run display is performed by an adult fairy-wren.

Coping with arid conditions

Physiological and behavioral adaptations

Emu in the Simpson Desert

The need to replenish body fluid can become acute in the hot and arid Australian outback. Humans cannot survive in the desert without a constant supply of drinking water, but over half of Australia's desert dwelling birds seldom drink. The remaining half depend on their skills in finding and conserving water. To survive in the arid environments, birds have adapted both physiologically as well as behaviorally.

One method for conserving water is to not excrete liquid wastes. To do this, many birds excrete uric acid in the form of a semi-solid paste. Also, Zebra Finches, Budgerigars, and Galahs are very efficient at absorbing water from their feces before defecating. Birds lacking this capacity must drink every day when air temperatures exceed 25°C (e.g., Crested Pigeons and Emus).

Maintaining a high body temperature somewhere between 40° C and 42° C can also help conserve water by reducing the need for evaporative cooling when ambient temperatures rise. In fact, most birds do not resort to panting until air temperatures reach an unbearable 45°C. Once a bird starts to pant, any further rise in temperature can be fatal. During the early 1930s, heat exhaustion—and not the lack of water—was the cause of death for thousands of desert dwelling Australian birds.

Normally, as body temperature rises, metabolic rate increases. This does not occur for some desert adapted birds such as Zebra Finches, Bourke's Parrots, Budgerigars and Spinifex Pigeons. All of these species have a relatively low metabolic rate that reduces the amount of food and water they need to survive.

Response to extreme heat

On January 5, 1983 at Shark Bay, Western Australia, midday temperatures on the ground reached 63°C while the coolest temperature in the shaded bush sizzled at 47°C with 22 percent relative humidity. Such temperatures far exceeded the physiological tolerance (45°C) of ground feeding birds. To escape the heat, one pair of White-browed Scrubwrens, four Splendid and five White-winged Fairy-wrens were found sheltering together in an underground hollow approximately 30 cm deep.

White-winged Fairy-wren

The entrance of the hollow was sheltered by an acacia bush, and near the bottom of the hollow the temperature hovered near 42°C (at 39.8 percent relative humidity). Both the Splendid Fairy-wrens and scrub-wrens left the hollow to feed, but neither species spent more than five minutes outside. None of the other birds left the protection of the undergrowth or left the vicinity of the hollow until late in the afternoon (6:00 pm) when outside

Splendid Fairy-wren

temperatures dropped to 37°C. Observers who found the birds noted a marked lack of aggression between the inhabitants of the hollow. Were the birds calm because it took too much energy to squabble?

Water holes

Physiological adaptations may not be enough to survive prolonged periods of hot, dry conditions. Under severe stress, many desert species leave in search of better conditions. The need for water is the principal reason Budgerigars, finches, and chats form large roaming flocks. In addition to their nomadic lifestyle, Zebra Finches and Budgerigars often travel as mated pairs. This arrangement allows such "rain adapted" species to start breeding quickly once they locate an area with water or appropriate food that can provide moisture.

The presence of a permanent water hole can support a thriving community of birds in which each species establishes a routine with regard to when, how much, and how often they drink. Some species prefer to drink early in the morning and again late in the afternoon, while other species visit water more or less regularly throughout the day. Finches, bowerbirds, magpies, honeyeaters, and Spinifex Pigeons are never far from water and must drink throughout the day. In contrast, White-browed Babblers, Yellow-rumped Thornbills, Variegated Fairy-wrens, Mistletoebirds, and shrike-thrushes seldom drink. Instead, they extract water from their food. Those species that visit water holes in large flocks include Crimson Chats, Crested Pigeons, Bourke's Parrots, Budgerigars, Zebra Finches, and occasionally Spiny-cheeked Honeyeaters.

Bishop

During summer, the pattern of early morning and afternoon visits to a water hole can minimize exposure to extreme midday temperatures and help avoid congestion at water holes. Indeed, the longer a bird must fly to reach water, the more important is its time of arrival. For instance, Bourke's Parrots congregate at water holes before sunrise or after sunset, Crested Pigeons, Ringnecks, and Mulga Parrots arrive early in the morning, while Galahs and Common Bronzewings prefer dusk.

Under the cover of darkness, Bourke's Parrots alight on the ground some 25–40 meters from water and then run in short bursts up to the edge to drink, as do Common Bronzewings. Budgerigars, on the other hand, often land in the middle of a pool and drink while floating. Budgies are so timid around water that a flock may vacate a pool soon after arriving, often before all members have had time to drink. Flock Bronzewings typically fly over a water hole in low circles before landing. When they decide it is safe they drop directly into the water and lie spread-eagle as they drink. By wetting their feathers they can carry some water back to their young. No matter how or when a water hole is approached, it is prudent to first survey the area for predators (e.g., Brown Goshawks, Collared Sparrowhawks, and a few falcons) that could be waiting in ambush.

Crimson Chat

Spinifex Pigeon

Spiny-cheeked Honeyeater

Diamond Doves (right) drink from the edge of a pool by submerging their bill and then pumping in water by movements of their throat and tongue. Budgerigars (left) arrive in large flocks and often land directly in the water. By dipping its beak into the water, a Budgerigar may drink its fill in two to four seconds.

Influence of diet

What a bird eats can determine how often it drinks. Seed and nectar feeders, for example, are all dependent on freestanding water (e.g., pigeons, parrots, and grass finches), whereas species that eat animal flesh such as insects can forego drinking even during the warmest and driest periods. There are two exceptions: aerial insectivores (e.g., woodswallows) and honeyeaters. Although birds in both groups consume insects, to catch their prey they must be active during the warmest times of the day. For this reason alone they must regularly drink water.

Masked Woodswallow

Drinking styles

There are 19 species of estrildid finches in Australia. All are small birds that specialize in eating seeds, and most live in semi-barren to grassy woodland habitat. Different species of finches can have different drinking styles. Some finches dip the bill into a pool of water and then tilt the

| Gouldian Finch | Pictorella Mannikin |

head upward to swallow the liquid collected in the lower mandible. To do this, however, requires that a bird sit close to and at approximately the same level as the water. Other species immerse the tip of the bill and suck the water without raising the head. When sucking, a bird can stretch the full length of its body to reach the water and can even hang upside down from a perch. To drink the same amount of water, swallowing is faster than sucking, although less versatile.

Both Pictorella Mannikins and Gouldian Finches are dependent on water holes scattered throughout northern Australia, but not to the same degree due to differences in how they drink—mannikins drink by swallowing and Gouldian Finches drink by sucking. Because mannikins spend less time quenching their thirst than Gouldian Finches, they are less vulnerable to predators (e.g., many birds of prey, kookaburras, butcherbirds, and feral cats). This difference gives Pictorella Mannikins an advantage over Gouldian Finches in regions where water is available in pools. In contrast, sucking comes in handy when water is less accessible. Gouldian

Double-barred Finches

139

Finches, unlike mannikins, can lean into cattle troughs or reach into rock crevices to get a drink. Shallow pools and even dewdrops can provide moisture to birds that can suck water.

Across northern Australia, mannikins range no farther west than eastern Kimberley; whereas Gouldian Finches occur in dry areas in central Kimberley and even farther west. Since the mid-1980s, three additional species—the Long-tailed, Double-barred, and Masked Finch—have extended their range, overlapping with the Gouldian Finch. This expansion was possible, in part, because all three species imbibe by sucking through their beaks.

Nocturnal birds

There is a behavioral strategy used by desert dwelling mammals that is rarely used by birds: namely, being active after dusk when temperatures are cooler. If we exclude owls and nightjars, which are active at night for reasons unrelated to excessive heat, there are only four desert dwelling species active at night in Australia. Of these, only the Night Parrot has shifted its lifestyle to escape diurnal heat stress. The other three have adopted nocturnal habits to exploit a particular food source: Letter-winged Kites eat nocturnal rodents; Inland Dotterels, which inhabit the clay pans and gibber plains of New South Wales, eat succulent plants during the day and insects at night; and Freckled Ducks, which visit ephemeral lakes and streams, consume both algae and aquatic invertebrates that are active only at night.

Inland Dotterel   Freckled Duck

140

Migration patterns

Erratic rainfall patterns are, in large part, responsible for much of the long-distance movements of birds in Australia, much more so than seasonal changes. In fact, it is inappropriate to label most species as migratory or nonmigratory since the same individual may be migratory, sedentary, or nomadic depending on its age (Tasmanian Silvereyes), geographical location, and current weather conditions (e.g., Eastern Spinebills).

An overview

Only 145 (out of 700) Australian species undertake large scale movements, and the majority of these species remain on the continent. Only a handful (e.g., cuckoos and some waterbirds) overwinter in New Guinea and islands farther north. Also, birds flying long-distances do not usually follow prominent flyways. For example, southern populations of Budgerigars and Rufous Whistlers migrate along an axis from southeast to northwest, whereas the Pallid Cuckoo and two species of songlarks (*Cinclorampus* spp.) follow circular routes that involve easterly movements northward in the autumn and, come spring, movement into the interior of the country.

It is not uncommon for individuals of a species residing in different regions of the country to move in different directions. Silvereyes breed-

ing in Tasmania, for example, seasonally migrate north across Bass Strait to the mainland and then fly northeast up the coast of Australia as far as southeast Queensland. In contrast, Silvereyes that breed in southeastern Victoria migrate to South Australia where they overwinter north of Adelaide. During some winters these separate populations of Silvereyes pass each other along the way.

To further complicate matters, some Australian species that exhibit seasonal movements are only partially migratory. Even

Silvereye eating Lantana

in Tasmania where winter conditions are harsh, only four species (Orange-bellied Parrot, Swift Parrot, Shining Bronze-Cuckoo, and Satin Flycatcher) are fully migratory out of some 20 that are known to have made it to the mainland. It is unknown why some individuals migrate while others remain in Tasmania throughout the year.

Transcontinental migration
Because Australasia is offset well east of the Eurasian landmass (see map below), any long-distance migration to Eurasia would require a very long flight over water with little chance of encountering favorable tail winds. In essence, long-distance movement between Australia and Southeast Asia would require migrants to make a 90 degree turn somewhere along the way. Only species such as shorebirds that are exquisitely adapted for long, nonstop flights migrate between Eurasia and Australia.

From the perspective of Asian migrants moving south, there is little to gain from a long strenuous flight to reach a continent that is overwhelmingly dominated by arid terrain. As a consequence, few Asian migrants winter in Australia, and the majority of those that do are shorebirds (30 species) that feed on intertidal mudflats and the wetland margins scattered across the continent.

Among the land birds, only 10 Asian species migrate to Australasia. Three species penetrate no farther than New Guinea, five reach the northern margin of Australia, and four are very uncommon. The remaining two species are swifts that reach the temperate areas of southern Australia.

Australian birds migrating farther north than Indonesia would have to make a sharp left turn to reach land. Also, the distance traveled, plus unfavorable winds, would make the trip prohibitive for most birds.

Types of migration
(see examples in table below)

Partial migration: Some, but not all, individuals in a local population migrate seasonally. Often an individual's age and sex determine whether it undergoes migration.

altitudinal migrants		*latitudinal migrants*	
Olive Whistler	Gang-gang Cockatoo	Yellow-faced & White-naped Honeyeater	
Flame Robin	Rufous Fantail	Welcome Swallow	Striated Pardalote

Obligate migration: All individuals seasonally migrate between breeding and overwintering sites.

international		*movement within the country*	
Rainbow Bee-eater	Common Koel	Swift Parrot (Tas.)	Spangled Drongo
Dollar Bird	Pallid Cuckoo		

nonbreeding visitors during the Australian summer			
Barn Swallow	Latham's Snipe	Common Tern	Fork-tailed Swift

Nomadism: Individuals that leave an area opportunistically in search of food, water, or breeding habitat. Approximately two-thirds of Australian species have populations that are nomadic.

Lorikeets	Flock Bronzewing	Stubble Quail	Crimson Chat
Emu	Chiming Wedgebill	Budgerigar	Letter-winged Kite

Vagrants: Species that breed in other countries but occasionally show up in Australia. Such vagrants are considered rare sightings.

American Golden-plover	Laughing Gull	Antarctic Tern
Black-backed Wagtail	Spotted Redshank	Red-throated Pipit

Useful facts

Parasitic cuckoos

In Australia, all but one species of cuckoo lay their eggs in the nests of other species. Among the remaining 11 species, there is stiff competition for hosts (see table below). The failure of Common Koels, for example, to exploit crows in Australia is probably due to the presence of the much larger Channel-billed Cuckoo that prefers to parasitize the nests of crows, ravens, Currawongs, and Australian Magpies. Common Koels parasitize Figbirds, Magpie-larks, and friarbirds. Perhaps for a similar reason, Pallid Cuckoos avoid large honeyeaters such as friarbirds in areas where the Common Koel breeds.

Horsfield's Bronze-Cuckoo

Cuckoos (weight)	Host species
Little Bronze-Cuckoo (17g)	gerygones
Shining Bronze-Cuckoo (23g)	Yellow-rumped and Brown Thornbills
Horsfield's Bronze-Cuckoo (23g)	fairy-wrens and thornbills
Black-eared Cuckoo (29g)	Speckled Warblers and Redthroats
Brush Cuckoo (36g)	honeyeaters with enclosed nests (N. Australia); fantails and robins
Fan-tailed Cuckoo (46g)	White-browed Scrubwrens, Brown Thornbills and Rock Warblers
Pallid Cuckoo (83g)	honeyeaters with open cup nests
Common Koel (225g)	Magpie-larks, Figbirds, friarbirds
Channel-billed Cuckoo (610g)	crows and currawongs

144

Juvenile Brush
Cuckoo (above)
is being fed by its
foster parent, a
Rufous Fantail.

Common Koel (male)

The world's largest obligate brood-parasite is the Channel-billed Cuckoo.

Corroborees

When territories are first established, various honeyeaters (e.g., Noisy
Miners and New Holland Honeyeaters) participate in communal displays comprised of three to 20 birds. In fact, the sight or sound of two
birds interacting will often prompt other members of the group, including both males and females, to join the corroboree.

During such brief gatherings, participants perform appeasement displays to one another, and as more birds join in, the spectacle becomes a
supernormal stimulus to nearby honeyeaters. Presumably interactions
during a corroboree help reduce aggression between members of the
group.

Research with New Holland Honeyeaters indicates that appeasement displays are only effective when participants recognize one another. If, for example, yellow wing feathers of a New Holland Honeyeater are covered, the bird will be attacked by members of its own group. Apparently the bird being attacked is not recognized as a New Holland Honeyeater. It is also important to be labeled as belonging to the same group. When a bird is removed from an aviary for several days and then reintroduced, it will promptly perform a submissive display that stimulates other honeyeaters to form a corroboree. This sequence of events, however, does not occur when a stranger is introduced into the group. More often, strangers are attacked.

There is no evidence that New Holland Honeyeaters individually recognize each other by sight. This was demonstrated when opaque contact lenses were fitted on captive birds, preventing them from discriminating plumage markings—listening to each other's calls was sufficient for the honeyeaters to recognize each other.

Novel source of water

Australian Pelicans normally obtain fresh water from the fish they eat. In Shark Bay Marine Park, however, an injured female was captured and nursed back to health by park rangers. While in captivity she learned to gulp fresh water from a sprinkler. She was eventually released, but later returned with several offspring to resume drinking at the sprinkler. The habit of gulping water from the sprinkler has now been passed down through three generations of pelicans.

Drawn from a Theo Allofs photo

146

Torpor in Tawny Frogmouths

It is well known that Tawny Frogmouths remain motionless for long periods, especially on winter days. Are they simply hiding from predators, or is there another reason? Some frogmouths in the southern states of Australia go into torpor to conserve energy.

Normally, a frogmouth's active body temperature ranges between 38° and 40° C, but during cool winter nights, its internal body temperature drops as low as 29° C. Come sunrise, its body temperature rises for a brief time before the bird moves to its daytime roost. Once settled, it may go into a second period of torpor lasting between one and three hours. As the bird passively basks in the sun, its body temperature gradually rises, reaching normal active levels by mid-afternoon.

Tawny Frogmouth

Torpor is usually undertaken at night when frogmouths are not feeding. Occasionally, however, when food supplies are difficult to come by, their activity and body temperatures will also fall. (Tawny Frogmouths feed on arthropods, a food source that is not plentiful during winter months.) The ability to go into torpor can help frogmouths endure short periods of food shortages. This, in turn, gives them the option of staying on their territory throughout the year.

At 500 grams, the Tawny Frogmouth is the largest known bird to go into torpor. There is evidence that Australian Owlet-nightjars also manipulate their body temperature to conserve energy stores.

Why some females are larger than males

For most species of birds, males and females are similar in size, and when there is a difference, males are usually larger than females. Yet, there are exceptions to this pattern. For example, females are larger than males in jacanas, jaegers, some shorebirds, woodpeckers, and many birds of prey (e.g., accipiters, falcons, hawks, eagles, and owls). What are the reasons that some female are larger than males? Several plausible explanations have been proposed.

Small males

In Peregrine Falcons, a female prefers a mate that can readily supply her and her offspring with food. In other words, she is looking for a male agile enough to capture highly maneuverable prey such as small birds. Presumably, a male's small size allows him to exploit a resource that is unavailable to a large female.

After comparing 12 species of falcon, one researcher found that those species in which males exclusively defend the territory were also species that exhibited the greatest differences in body size between males and females. Here again, small size promotes agility that can translate into more impressive aerial displays.

A male's small size may also enhance his survival very early in life while growing up in the nest. It is interesting to note that in many raptors males develop more rapidly than females, and hence, fledge earlier than their sisters. This is a clear advantage when there is a real danger of being killed by a larger sibling. Siblicide is common in many birds of prey including falcons.

Among Grey Goshawks, the male (left) is less than half the weight of the female (right). Females also have a relatively larger beak and talons than males. (Drawn from a photograph in Olsen 1995).

148

In the Peregrine Falcon, females are noticeably larger than males, as seen when both sexes visit the nest site.

Large females

Most females gain weight before and during egg laying, presumably to increase their energy reserves during a period when physiological stress is high. Naturally, large females can store more reserves than small females. These reserves could enhance a bird's survival during short periods of famine. Furthermore, since female raptors spend more time at the nest than males, they are more likely than males to be called upon to defend offspring from predators. In this context, large size can be very beneficial. And lastly, during the non-breeding season, large females usually defend the best feeding sites.

Birds and willy-willies

Warm conditions across the interior of Australia are favorable for the creation of hot spots on the ground that warm the air. As the hot air rises, a pocket of low pressure is left behind causing cooler air to spiral in and fill the void. If a light breeze is blowing, the rising vortex can break away to become a dust devil or a willy-willy. These swirling vortexes can move across the landscape at speeds up to 15 kilometers per hour and generate internal winds of 80 kilometers per hour. A large willy-willy is capable of lifting dust, grass, and birds to heights of several hundred meters.

While Australian Hobbies, kestrels, and even large Wedge-tailed Eagles make use of thermals to gain altitude, flocks of Galahs seem to have another purpose in mind. When Galahs enter a willy-willy, they flap their wings and screech. When they are thrown out, they abruptly turn and re-enter at the base of the vortex to repeat the process over and over. Most reports of Galahs playing in willy-willies originate in the districts of Coonabarabran and Tamworth of northern New South Wales. This regional bias suggests that a few local birds have learned on their own how to ride in willy-willies. It is assumed that the practice is culturally transmitted to other Galahs in the local population.

Crows (above) and bee-eaters (below) are known to play with small stones.

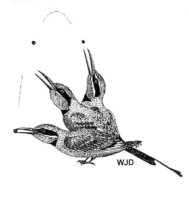

Australian Magpies are at play.

Do birds play?

Play-like behavior tends to defy explanation. Consider the reported antics of a pair of Torresian Crows. One of the birds retrieved and flew off with a yellow golf ball in its bill, only to drop the ball on the nearby pavement. Both crows pursued and recovered the bouncing ball only to repeat the performance. Since the birds actually took turns carrying the ball, it would seem as if the birds were playing.

Play often involves mock fighting in slow motion, with blows packing very little punch. Most telling, participants in play fights usually take turn playing the attacker. While playing, a pair of magpies may indulge in mock fights, aerial chases, or hang from the branches of trees. Elements of sexual behavior are often exhibited when groups of magpies play with one another. If you watch long enough, a number of disconnected behaviors are performed, but not in any discernible pattern.

Three common types of play exhibited by birds are object manipulation, locomotor play, and social play (which involves interactions between two or more individuals).

To tell if birds are playing, look for an absence of aggression. For example, aggressive displays are not used during mock fights. Also, play is not undertaken to achieve any specific goal. And lastly, play can be highly contagious among members of a group.

150

In New South Wales, a White-winged Chough was observed lying on its side playing with a pine cone. It kicked and scratched at the cone the way a kitten plays with a small ball. While lying on its back, it used one foot to hold the cone and then tore at it with the other foot. A similar scenario was documented with Pied Butcherbirds and Little Corellas. Nestlings indulge in locomotor play of sorts when they vigorously flap their wings while holding tightly onto the edge of the nest. Once able to fly, young birds of many species perform fancy aerial maneuvers that seem to serve no practical function. Such acrobatic flights are typical of raptors and large flocking birds such as gulls, crows, and parrots.

Time to play
Most young birds go through sensitive periods during which special tasks are learned such as singing, handling social relations, negotiating difficult flight maneuvers, and fighting. Such activities require well coordinated neural-motor patterns. Play provides an opportunity for young birds to practice and learn critical activities at a time in their life when the brain and motor skills are developing.

Tool use
There are 30 or more species of birds that use tools. Some birds throw rocks at eggs to break them (e.g., Black-breasted Buzzards). Other species use sticks and bark to extract insects from crevices (e.g., Varied Sittellas). Australian Butcherbirds routinely anchor their prey on spines or between the notch formed by a broken branch. Dropping prey is akin to throwing stones. Seagulls and crows, for example, drop mollusks to break open the hard shell.

On another note, a frequently cited example of avian tool use is bait-fishing by Striated Herons in Australia. Some herons select a small object (twig, leaf, berry, feather, insect, earthworm, bread, cracker, piece of foam, etc.) and drop it into the water. The bird then waits in ambush until fish come to investigate the "bait."

SD

Black-breasted Buzzard hits a rock against an egg to open it.

151

In the wild, a few herons have been observed using bait to catch fish. But there is no evidence that the herons learn the technique from each other. So what is going on? It appears that juvenile herons, while manipulating objects, will on occasion drop them in the water, an event that may attract the attention of fish looking for food. In a few of these cases, the young herons have learned to associate dropping items in the water with the arrival of fish. Such early experiences are necessary to acquire the skill of fishing with bait. However, since adult herons do not play with objects, it is unlikely that birds can acquire the knowledge that fish can be enticed to the surface with bait. Neither can they learn such skills by watching other herons.

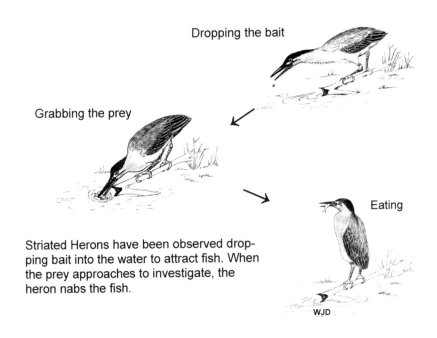

Dropping the bait

Grabbing the prey

Eating

Striated Herons have been observed dropping bait into the water to attract fish. When the prey approaches to investigate, the heron nabs the fish.

WJD

Keeping tally

"Yeah, and then she says she likes smaller males..."

Indices* of sexual size differences in raptors (from Olsen 1995)

Species	Weight*	Wing length	Main prey (secondary prey)
Grey Goshawk	1.9*	1.20	birds (mammals)
Collared Sparrowhawk	1.8	1.16	small passerines (insects)
Peregrine Falcon	1.5	1.16	birds
Osprey	1.4	1.08	fish (birds)
Brahminy Kite	1.2	1.04	insects (invertebrates)
Pacific Baza	1.15	1.03	insects (tree dwelling frogs and lizards)
Nankeen Kestrel	1.1	1.04	insects & birds (rodents)

* Each index expresses how much larger the female is compared to the male. For example, the female Grey Goshawk is nearly twice as heavy as the male; to be more precise, she is 1.9x heavier.

Nonvocal sounds used by Australian birds

Family	Source of sound
Non-perching birds	
Grebes	bill-fencing
Penguins	bill-fencing
Albatrosses	bill-snapping and billing
Cormorants	vibration of throat pouch
Frigatebirds	bill-clacking amplified by pouch
Herons	bill-snapping and fencing
Musk Ducks	wings slapping water
Whistling Ducks	wing feathers in flight
Snipe	tail feathers during display
Pigeons	wing primaries, wing clapping
Cockatoos	feet, drumming with object
Owls	bill-snapping
Kingfishers	bill-snapping
Frogmouths	bill-snapping
Swifts	wing feathers in flight
Perching birds	
Lyrebirds	feet, shaking bushes
Rock Warblers, etc.	wing feathers during displays
Chowchillas	wing feathers during displays
Fantails, Magpie-larks	bill-snapping
Honeyeaters	bill-snapping
Butcherbirds, etc.	bill-snapping
Birds of Paradise	body and wing feathers when displaying on a perch

Mouth color of Australian birds

Some species, adults, have markings (dots, stripes) on the tongue and palates; within a species, mouth color can vary geographically. Colors are from photographs published in the popular literature.

Common Name	Internal	Flanges	Comments
Dollar Bird	pale white	pale white	hole-nester
Owlet-nightjar	white to pale	pink cream	hole-nester
Pheasant Coucal	bright red	black	red tongue lined with black; also two black spots on palate
Lewin's Honeyeater	yellow	yellow	open nest
Noisy Miner	orange	yellow	open nest
New Holland Honeyeater	jet black	bright yellow	open nest
Pale-yellow Robin	red	yellow	open nest
Mistletoebird	bright red	yellow	dome nest
Rufous Fantail	red-orange	yellow	open nest
Crested Bellbird	bright yellow	yellow	open nest
Crimson Chat	pale yellow	yellow	open nest
Brush Cuckoo	pale yellow	pale yellow	brood parasite

Drinking habits of desert dwelling birds

Families	Needs to drink?	Diet *
Emus	yes	mixed
Kites, eagles, and harriers	yes & no	flesh
Falcons	no	flesh
Pigeons, doves	yes	seeds
Cockatoos, corellas, etc.	yes	seeds, fruit
Parrots, lorikeets, etc.	yes	seeds, fruit
Cuckoos	no	insects
Nightjars	yes	insects
Kingfishers	no	carnivorous
Bee-eaters	no	insects
Singing bushlarks, skylarks	no	insects
Fairy-wrens, grasswrens, etc.	no	insects
Cuckoo-shrikes, and trillers	yes (occ.)	insects
Whistlers, shrike-thrushes, etc.	no (occ.)	insects
Babblers	yes & no	insects
Thornbills, pardalotes, etc.	no	insects
Swallows and martins	yes	insects
Honeyeaters, chats	yes	nectar, insects
Grassbirds, songlark, etc.	no	insects
Butcherbirds, etc.	yes	insects, etc.
Bowerbirds	yes	mixed
Ravens and crows	yes	mixed

* During summer
"mixed" indicates omnivorous diet
"occ." indicates occasional drinking

Adaptations of desert dwelling birds

Physiological

Uric acid excretion

High body temperature

Ability to lower metabolic rate in summer, e.g., Zebra Finches, Spinifex Pigeon, some parrots, and scrubwrens.

Enhanced absorption of water in digestive tract, e.g., Galahs and other desert dwelling birds.

Tolerance of salt in bloodstream allowing birds to drink salty water, e.g., Zebra Finch.

Ability to eliminate salt via nasal salt secretion, e.g., some raptors.

Behavioral

Nomadism: unpredictable and irregular movements in response to local weather and food supply, e.g., Budgerigars, Emu, bronzewings, and Stubble Quail.

Adjusts diet.

Adjusts drinking schedules at water holes (see page 137—139 for details).

Shifts to crepuscular and nocturnal lifestyle, e.g., Night Parrot

Seeks shelter in caves, under rocks, and in burrows.

Foraging innovations

Species	Innovation
White-winged Chough	uses an empty mussel shell to hammer to open shells of live mussels
Pacific Gull	drops mussels onto hard surfaces
Crimson Rosella	sallies for insects on the wing
Black Kite	uses bread as bait to catch fish
Shrike-tit	breaks off a twig to use as tool to pry out prey from bark of trees
Grey Shrike-thrush	uses a twig to probe crevices in house brick
Pied Butcherbird	inserts strips of meat through clothesline fastener to soften the meat
Torresian Crow	turns over dead Cane Toads to eat them through the underbelly to avoid the dorsal poison glands

Vocal mimicry

Approximately 15 percent of Australian songbirds copy sounds, usually those of other species. Most of these vocal mimics occur in scrub, heath, or in forest habitats where visibility is relatively limited. The most accomplished mimics are not usually brightly colored. The better known mimics, excluding parrots, are listed below (based on Chisholm 1950).

Major mimics

Lyrebirds (Superb and Albert's)
Bowerbirds (Satin, Western, and Great)
Chestnut-rumped Heathwren
Rufous Scrub-bird
Pied Butcherbird
Redthroat
Yellow-throated Scrubwren
Brown Songlark
Australian Magpie

Minor mimics

Orioles (Olive-backed and Yellow)
Grey Butcherbird
Bowerbirds (Fawn-breasted
             and Golden bowerbirds)
Logrunner
Thornbills (most species)
Whistlers (Olive and Golden)

Introduced mimics

European Blackbird
Common Myna
European Starling
Skylark
Song Thrush

"You realize, none of that is his own stuff!"

Notes

References

Species accounts — Behavioral profiles

Albert's Lyrebird (*Menura alberti*)

Baylis, J.R. 1982. Avian vocal mimicry: its function and evolution. p. 51–83. In: *Acoustic Communication in Birds*, Vol. 2. Eds. Kroodsma, D.E. & Miller, E.H. Academic Press, New York.

Curtis, H.S. 1972. The Albert Lyrebird in display. *Emu* 72:8–84.

Reilly P.N. 1988. *The Lyrebird*. New South Wales University Press, Sydney.

Robinson, F.N. 1977. Environmental origin of the menurae. *Emu* 77:167–168.

Robinson, F.N. & Curtis, H.S. 1996. The vocal displays of the lyrebirds (Menuridae). *Emu* 96:258–275.

Australian Brush-turkey (*Alectura lathami*)

Curio, E. 1993. Proximate and developmental aspect of antipredator behavior. *Adv. Study Behav.* 22:135–238.

Göth, A. 2001. Innate predator-recognition in Australian Brush-turkey (*Alectura lathami, Megapodiidae*) hatchlings. *Behaviour* 138:117–136.

Jones, D.N. 1990. Social organisation and sexual interactions in Australian Brush-turkeys (*Alectura lathami*): implications of promiscuity in a mound-building megapode. *Ethology* 84:89–104.

Jones, D.N 1990. Male mating tactics in a promiscuous megapode: patterns of incubation mound ownership. *Behav. Ecol.* 1:107–115.

Jones, D.N. et al. 1995. The Megapodes. Oxford University Press, Oxford.

Schaller, G.G. et al. 1962. The ontogeny of avoidance behaviour in some precocial birds. *Anim. Behav.* 370–381.

Australian Magpie (*Gymnorhina tibicen*)

Brown, E.D. & Veltman, C.J. 1987. Ethogram of the Australian Magpie in comparison to other cracticidae and corvus species. *Ethology* 76:309–333.

Brown, E.D. & Farabaugh, S.M. 1991. Macrogeographic variation in alarm calls of the Australian Magpie. *Bird Behaviour* 9:64–68.

Farabaugh, S.M. et al. 1988. Song sharing in a group-living songbird, the Australian Magpie. Part II. Vocal sharing between territorial neighbors, within and between geographical regions, and between sexes. *Ethology* 104:105–125.

Floyd, R.B & Woodland, D.J. 1981. Localization of soil dwelling scarab larvae by the Black-backed Magpie. *Anim. Behav.* 29:510–517.

Hughes, J.M et al.1983. Territories of the Australian Magpie *Gymnorhina tibicen* in southeast Queensland. *Emu* 83:108–111.

Kaplan, G. 2004. *Australian Magpie: Biology and Behaviour of an Unusual Songbird.* Australian Natural History Series, CSIRO, Vic.

Nealson, T. 2000. Parental care in aggressive and non-aggressive Australian Magpies: An experimental approach. Honours dissertation. Griffith University, Brisbane.

Sanderson, K. & Crouch. 1993. Vocal repertoire of the Australian Magpie in South Australia, *Aust. Bird Watcher* 15:162–164.

Australian Pelican (*Pelecanus conspicillatus*)

Crawford, M. 1987. Predation of Grey Teal by a Pelican. *Bird Obs.* 663:47.

Davies, R. 1986. Driven to Drink. *Bird Obs.* 10: 648–649.

Marchant S. & Higgins P.J. Eds. *Handbook of Australian, New Zealand and Antarctic Birds.* Vol. 1 Ratites to Ducks. Oxford University Press. 1990.

Thomas D.A. 1986. Pelicans. *Bird Obs.* 655:80.

Australian Wood Duck (*Chenonetta jubata*)

Briggs, S.V.1991. Effects of egg manipulations on clutch size in Australian Wood Ducks. *Emu* 91:230–235.

Frith, H.J. 1982. *Waterfowl in Australia.* Angus and Robertson. Sydney. Australia.

Johnsgard, P.A. 1965. *Handbook of Waterfowl Behaviour.* Constable & Co. Ltd. London.

Kingsford, R.T. 1989. Food of the Maned Duck *Chenonetta jubata* during the breeding season. *Emu* 89:119–124.

Bell Miner (*Manorina melanophrys*)

Higgins, P.J., Peter, J.M. & Steele, W.K. Eds. 2001. *Handbook of Australian, New Zealand and Antarctic Birds. Volume 5. Tyrant-Flycatchers to Chats.* Melbourne: Oxford University Press.

Budgerigar (*Melopsittacus undulatus*)

Higgins, P.J. Ed. 1999. *Handbook of Australian, New Zealand and Antarctic Birds. Volume 4. Parrots to Dollarbird.* Melbourne: Oxford University Press.

Cassowary (*Casuarius casuarius*)

Mack, A.L. & Jones, J. 2003. Low frequency vocalizations by cassowaries (*Casuarius*). *Auk* 120:1062-1068.

Cattle Egret (*Ardea ibis*)

Baker, D. 1969. *Ostrich* 40:75–129.

Jenni, D.A. 1969. *Ecol. Monogr.* 39:245–249.

Marchant S & Higgins PJ Eds. 1990. *Handbook of Australian, New Zealand and Antarctic Birds,* Vol.1, Pt B, Australian Pelicans to Ducks, Melbourne OUP :1017–1028.

McKilligan NG 1984. The food and feeding ecology of the Cattle Egret, *Ardea ibis,* when nesting in south-east Queensland. *Aust.Wildl.Res.* 11:133–144.

Channel-billed Cuckoo (*Scythrops novaehollandiae*)

Higgins, P.J. Ed. 1999. *Handbook of Australian, New Zealand and Antarctic Birds. Volume 4. Parrots to Dollarbird.* Melbourne: Oxford University Press.

Common Koel (*Eudynamys scolopacea*)

Higgins, P.J. Ed. 1999. *Handbook of Australian, New Zealand and Antarctic Birds. Volume 4. Parrots to Dollarbird.* Melbourne: Oxford University Press.

Maller, C.J. and Jones, D. N. 2001. Vocal behaviour of the Common Koel, *Eudynamys scolopacea,* and implications for mating systems. *Emu* 101: 105-112.

Comb-crested Jacana (*Irediparra gallinacea*)

Demong, N.J & Emlen, S.T. 1995. Signals of the flesh. *Nature Australia,* Spring issue.

Emlen, S.T. et al. 1998. Seduced and abandoned single parent birds suffer female infidelity, then get the job of bringing up baby. Cornell News.

Jenni, D.A. & Collier, G. 1972. Polyandry in the American Jacana. *Auk* :89 743–765.

Mace, T.R. 2000. Time budgets and pair bond dynamics in the Comb-crested Jacana (*Irediparra gallinacea*): A test of a hypothesis. *Emu* 100:31–41.

Marchant, S. & Higgins, P.J. Eds. *Handbook of Australian, New Zealand and Antarctic Birds.* Vol.2: Raptors to Lapwings. Oxford University Press, Oxford, 1993.

Crested Pigeon (*Ocyphaps lophotes*)

Baldwin, M. 1976. Sunbird 7:59–64.

Frith, H.J 1977. Some display postures of Australian pigeons. *Ibis* 119:167–182.

Goodwin, D. 1970. *Pigeons and Doves of the World*. The British Museum. England.

Higgins, P.J. & Davies, S.J.J.F. Eds. 1996. *Handbook of Australian, New Zealand & Antarctic Birds*: Volume Snipe to Pigeons. Oxford University, Press, Oxford.

Darter (*Anhinga melanogaster*)

Frederick, P.C. & Siegel-Causey, D. 2000. *The Birds of North America*. No. 522.

Hustler, K. 1992. Buoyancy and its constraints on the underwater behavior of Reed Cormorants (*Phalacrocorax africanus*) and Darters (*Anhinga melanogaster*). *Ibis* 132: 229–236.

Marchant, S. & Higgins, P.J. (eds.) 1990. *Handbook of Australian, New Zealand and Antarctic Birds Vol.1 Part 2* Australian Pelicans to Ducks. O.U.P. Melbourne: Oxford University Press.

Vestjens, W.J.M. 1975. Breeding behaviour of Darters at Lake Cowal, N.S.W. *Emu* 75: 121–131.

Dusky Moorhen (*Gallinula tenebrosa*) **and other rails**

Alvarez, F. 1993. Alertness signaling in two rail species. *Anim. Behav.* 46:1229–1231.

Garnett, S.T. 1978. The Behaviour Patterns of the Dusky Moorhen, *Gallinula tenebrosa*. *Aust. Wildl. Res.* 7:102–112.

Goldizen, A 1998. Personal Communication.

Ryan, D.A., et al. 1996. Scanning and tail-flicking in the Australian Dusky Moorhen (*Gallinula tenebrosa*). *Auk* 113:499–501.

Eastern Whipbird (*Psophodes olivaceus*)

Higgins, P.J. & Peter, J.M. Eds. 2002. *Handbook of Australian, New Zealand and Antarctic Birds. Volume 6. Pardalotes to Spangled Drongo*. Melbourne: Oxford University Press.

Rogers, A.C. et al. 2006. Duet duels: sex differences in song matching in duetting Eastern Whipbirds. *Anim. Behav.* 72: 53-61.

Eclectus Parrot (*Eclectus roratus*)

Heinsohn, R. & Legge, S. 2001. Seeing red: a parrot's perspective. *Nature Australia*, Winter, 34–40.

Higgins, P. J. (ed.) 1999. *Handbook of Australian, New Zealand and Antarctic Birds. Vol.* 4.Oxford University Press, Melbourne.

Legge. S. et al. 2004. The availability of nest and breeding population size of Eclectus Parrots, *Eclectus roratus*, on Cape York Peninsula, Australia. *Wildl. Res.* 31:149-151.

Emu (*Dromaius novaehollandiae*)

Blache, D. et al. 2000. Social mating system and sexual behaviour in the Emu, (*Dromaius novaehollandiae*). *Emu* 100:161–168.

Blache, D., et al. 2000. Seasonality in Emus (*Dromaius novaehollandiae*). In: *Avian Endocrinology*. Eds: A Dawson and Chaturvedi, C.M., Narosa Publishing House, New Dehli. Pp. 129–139.

Coddington, C.L. & Cockburn, A. 1995. The mating system of free-living Emus. *Australian Journal of Zoology* 43:365–372.

Davies, S.J.J.F. 1967. Sexual dimorphism in the Emu. *Emu* 67:23–26.

Davies, S.J.J.F. 1984. Nomadism as a response to desert conditions in Australia. *Journal of Arid Environments* 7:183–195.

Davies, S.J.J.F. et al. 1971. The results of banding 154 Emus in Western Australia. *C.S.I.R.O. Wildlife Research* 16:77–79.

Gaukrodger, D.W. 1925. The Emu at home. *Emu* 25:53–57.

Malecki, I.A. et al. 1996. Semen production by the male Emu (*Dromaius novaehollandiae*). 1. Methods for collection of semen. *Poultry Science* 76:615–621.

Schrader, N. 1975. Emu incubating Paddy-melons. *Emu* 75:43.

Taylor, E.L. et al. 2000. Genetic evidence for mixed parentage in nests of the Emu (*Dromaius novaehollandiae*). *Behavioral Ecology & Sociobiology* 47:359–364.

Galah (*Cacatua roseicapilla*) & **Major Mitchell** (*Cacatua leadbeateri*)

Higgins P.J. Ed. *Handbook of Australian, New Zealand & Antarctic Birds.* 1999 Vol. 4 :104–163.

Rogers, L.J. & McCulloch, H. 1981. Pair-bonding in the Galah (*Cacatua roseicapilla*). *Bird Behav.* 3: 80–92.

Rowley, I. & Chapman, G. 1986. Cross-fostering, imprinting, and learning in two sympatric species of cockatoo. *Behaviour* 96:1–16.

Rowley, I. 1990. *Behavioural Ecology of the Galah.* Surrey Beatty & Sons, Australia.

Westcott, D.A. & Cockburn, A. 1988. Flock size and vigilance in parrots. *Aust. J. Zool.* 55: 335–349.

Grey Butcherbird (*Cracticus torquatus*)

Grey-crowned Babbler (*Pomatostomus temporalis*)

King, B.R. 1974. Communal Breeding and Related Behaviour in the Grey-crowned Babbler in Southeast Queensland. Master's Thesis. Zoology Department, Queensland University.

King, B.R. 1980. Social Organization and Behaviour of The Grey-Crowned Babbler (*Pomatostomus temporalis*). *Emu* 80:59–76.

Hoary-headed Grebe (*Poliocephalus poliocephalus*)

Fjeldsa, J. 1983. Social behaviour and displays of the Hoary-headed Grebe *P. poliocephalus*. *Emu* 3:129–139.

Laughing Kookaburra (*Dacelo novaeguineae*)

Heinz-Ulrich, R. & Schmidl, D. 1988. Helpers have little to laugh about: group structure and vocalisation in the Laughing Kookaburra (*Dacelo novaeguineae*). *Emu* 88:150–160.

Legge, S. 2004. *Kookaburra: King of the Bush.* CSIRO, Australia.

Parry, V.A. 1970. Kookaburras. Landsdowne Press. Australia.

Lewin's Honeyeater (*Meliphaga lewinii*)

Higgins, P.J., Peter, J.M. & Steele, W.K. Eds. 2001. *Handbook of Australian, New Zealand and Antarctic Birds. Volume 5. Tyrant-Flycatchers to Chats.* Melbourne: Oxford University Press.

Little Bronze-cuckoo (*Chrysococcyx minutillus*)

Higgins, P.J. Ed. 1999. *Handbook of Australian, New Zealand and Antarctic Birds. Volume 4. Parrots to Dollarbird.* Melbourne: Oxford University Press.

Little Penguin (*Eudyptula minor*)

Cannell, B.L. & Cullen, J.M. 1998. The foraging behaviour of Little Penguins (*Eudyptula minor*) at different light levels. *Ibis* 140:467–471.

Chiaradia, A.F. 1999. Breeding biology and feeding ecology of Little Penguins (*Eudyptula minor*) at Phillip Island—a basis for a monitoring program. Ph.D. thesis. University of Tasmania.

Fadely, J. 1991. Vocal recognition by Little Penguins, M.Sc. Thesis. *San Jose State University*

Stahel, C.D. et al. 1984. Sleep and metabolic rate in the Little Penguin, *Eudyptula minor*. *Journal of Comparative Physiology B* 154:487–494.

Stahel, C. & Gales, R. 1987. *Little Penguin, Fairy Penguins in Australia*. Australian Natural History Series, ed. Dawson, T. *New South Wales University Press*.

New Holland Honeyeater (*Phylidonyris novaehollandiae*)

Armstrong, D.P. 1996. Territorial behaviour of breeding White-cheeked and New Holland Honeyeaters: conspicuous behaviour does not reflect aggressiveness. *Emu* 96:1–11.

McFarland, D.C. 1996. Aggression and nectar use in territorial non-breeding New Holland Honeyeaters (*Phylidonyris novaehollandiae*) in eastern Australia. *Emu* 96:181–188.

Rooke, I.J. 1979. The social behaviour of the New Holland Honeyeater (*Phylidonyris novaehollandiae*) Ph. D. Thesis, University of Western Australia, Perth: RAOU microfiche No. 3, Melbourne.

Noisy Miner (*Manorina melanocephala*)

Dow, D.D. 1975. Displays of the Honeyeater (*Manorina melancephala*). *Z. Tierpsychol.* 38:70–96.

Dow, D.D. & Whitmore, M.J. 1990. Noisy Miners: Variation on the theme of communality. In. *Cooperative Breeding in Birds* Eds. Stacey, P.B. & Koenig, P.B. Cambridge University Press. Cambridge.

Põldmaa, T. & Holder, K. 1997. Behavioural correlates of monogamy in the Noisy Miner, (*Manorina melanocephala*). *Anim. Behav.* 54:571–578.

Magpie-lark (*Grallina cyanoleuca*)

Hall, M.L. & Magrath, R.D. 2000. Duetting and mateguarding in Australian Magpie-larks (*Grallina cyanoleuca*). *Behav. Ecol. Sociobiol.* 47:180–187.

Masked Lapwing (*Vanellus miles*)

Marchant, S. & Higgins, P.J. Eds. 1999. *Handbook of Australian, New Zealand & Antarctic Birds*. Vol. 2:943–957.

Mistletoebird (*Dicaeum hirundinaceum*)

Reid, N. 1997. The Mistletoebird and Australian Mistletoes: Co-evolution or coincidence? *Emu* 87: 130-131.

Pheasant Coucal (*Centropus phasianinus*)

Higgins, P.J. Ed. 1999. *Handbook of Australian, New Zealand & Antarctic Birds*. Vol. 4:793–806.

Pied Butcherbird (*Cracticus nigrogularis*)

Rainbow Lorikeet (*Trichoglossus haematodus*)

Serpell, J. 1981. Duets, greetings and triumph ceremonies: analogous displays in the Parrot genus *Trichoglossus*. *Z. Tierpsychol.* 55:268–283.

Higgins, P.J. Ed. Handbook of Australian, New Zealand & Antarctic Birds. 1999, Vol. 4:195–212.

Red-rumped Parrot (*Psephotus haematonotus*)

Regent Honeyeater (*Xanthomyza phrygia*)

Franklin, D. & Menkhorst, P. 1988. The bare facial skin of the Regent Honeyeater. *Australian Bird Watcher* 12: 237–238

Geering, D.G. & French, K. 1998. Breeding biology of the Regent Honeyeater (*Xanthomyza Phrygia*) in the Capertee Valley, New South Wales. *Emu* 98: 104–116.

Veerman, P.A. 1991. Vocal mimicry of larger honeyeaters by the Regent Honeyeater (*Xanthomyza Phrygia*). *Australian Bird Watcher* 14:180–189.

Veerman, P.A. 1991. Batesian acoustic mimicry by the Regent Honeyeater (*Xanthomyza Phrygia*). *Australian Bird Watcher* 15: 250–259.

White, H.L. 1909. Warty-faced honeyeaters and friarbirds. *Emu* 9:93–94

Restless Flycatcher (*Myiagra inquieta*)

Rock Dove (*Columba livia*)

Fabricius, E. & Jansson, A 1963. Reproductive behavior of the pigeon (*Columbia livia*). *Anima. Behav.* 11:534-47.

Goodwin, D. 1967. *Pigeons and Doves of the World.* London: British Museum.

Higgins, P.J. & Davies, S.J.J.F. Eds. 1996. *Handbook of Australian, New Zealand and Antarctic Birds. Volume 3. Snipe to Pigeons.* Melbourne: Oxford University Press.

Satin Bowerbird (*Ptilonoryhynchus violaceus*)

Borgia, G. 1985 Bower quality, number of decorations and mating success of male Satin Bowerbirds (*Ptilonorhychus violaceus*): an experimental analysis. *Anim. Behav.* 33:266–271.

Borgia, G. 1985. Bower destruction and sexual competition in the Satin Bowerbird (*Ptilonorhychus violaceus*). *Behav. Ecol. and Sociobiol.* 18:91–100.

Borgia, G. 1986. Sexual selection in bowerbirds. *Sci. Amer* 254:92–100.

Borgia, G. & Gore, M.A. 1986. Feather stealing in the Satin Bowerbird (*Ptilonorhychus violaceus*): male competition and the quality of display. *Anim. Behav.* 34:727–738.

Coleman, S.W., et al. 2004. Variable female preferences drive complex male displays. *Nature* 428:742–745. Also presented at *ABS* conference.

Doucet, S.M. & Montgomerie, R. 2003. Bower location and orientation in Satin Bowerbird: optimising the conspicuousness of male display? *Emu* 103:105–109.

Hunter, C.P. & Dwyer, P.D. 1997. The value of objects to Satin Bowerbirds (*Ptilonorhychus violaceus*). *Emu* 97:200–206.

Loffredo, C.A. & Borgia, G. 1986. Male courtship vocalization as cues for mate choice in the Satin Bowerbird (*Ptilonorhynchus violaceus*). *Auk* 103:189–195.

Marshall, A.J. 1954. *Bowerbirds: Their displays and breeding cycle.* Oxford University Press.

Patricelli, G.L., et al. 2002. Sexual selection: male displays adjust to female's response. *Nature* 415:279–280.

Patricelli, G.L., et al. 2004. Female signals enhance the efficiency of mate assessment in Satin Bowerbirds (*Ptilonorhynchus violaceus*). *Behav. Ecol.* 15:297–304.

Procter-Gray, E. & Holmes, R.T. 1981. Adaptive significance of delayed attainment of plumage in male American Redstarts: Tests of two hypotheses. *Evolution* 35:742–751.

Rhijn, J.G. van 1973. Behavioural dimorphism in male Ruffs (*Philomachus pugnax*) L. *Behav.* 47:153–229

Thompson, C.W. 1991. The sequence of molts and plumages in painted buntings and implications for theories of delayed plumage maturation. *Condor* 93:209–235.

Vellenga, R. 1970. Behaviour of the male Satin Bowerbird at the bower. *Australian Bird Bander* 1:3–11.

Sulphur-crested Cockatoo (*Cacatua galerita*)

Striated Pardalote (*Pardalotus striatus*)

Higgins, P.J. & Peters, J.M. Eds. *Handbook of Australian, New Zealand & Antarctic Birds.* Vol. 6 .

Woinarski, J.C.Z. 1984. Small birds, lerp-feeding and the problem of honeyeaters. *Emu* 84:137–141.

Superb Fairy-wren (*Malurus cyaneus*) and relatives

Langmore, N.E. & Mulder, R.A. 1992. A novel context for bird song: predator calls prompt male singing in the kleptogamous Superb Fairy-wren, (*Malurus cyaneus*). *Ethology* 90:143–153.

Mulder, R.A. et al. 1994. Helpers liberate female Fairy-wren from constraints on extra-pair mate choice. *Proc. R. Soc. Lond. B* 255:223–229.

Mulder. R.A. 1997. Extra-group courtship displays and other reproductive tactics of Superb Fairy-wrens. *Aust. J. Zool.* 45:131–143.

Pruett-Jones S.G. & Lewis M.L. Sex ratio and habitat limitation promote delay dispersal in Superb Fairy-wrens. *Nature* 348: 541–542

Rowley, I. 1962. Rodent-run distraction display by a passerine the Superb BlueWren (*Malurus cyaneus*) (L.). *Behaviour* 19:170–176.

Rowley, I. 1965. The life history of the Superb Blue Wren (*Malurus cyaneus*). *Emu* 64: 251–297.

Rowley, I. 1991. Petal-carrying by fairy-wrens of the genus *Malurus.* *Bird Watcher* 14: 75–81.

Willie Wagtail (*Rhipidura leucophrys*)

Cameron, E. 1985. Habitat usage and foraging behaviour of three fantails (*Rhipidura: pachycephalidae*). In: *Birds of Eucalypt Forest and Woodlands: Ecology, Conservation, Management*, Eds. Keast, A., Recher, H.F., Ford, H. & Saunders, D. RAOU and Surrey Beatty, Sydney

Clapp, G.E. 1982. Notes on the behaviour of the Willie Wagtail *Rhipidura leucophrys* in Papua New Guinea. *Papua New Guinea Bird Society Newsletter* 187–188.

Dyrez, A 1994. Breeding biology and behaviour of the Willie Wagtail (*Rhipidura leucophrys*) in the Madang Region, Papua New Guinea. *Emu* 94: 17–26.

Goodey, W. & Lill, A. 1993. Parental care by the Willie Wagtail in southern Victoria. *Emu* 93: 180–187.

Hough, K. 1968. Courtship display by Willie Wagtail. *Emu* 68: 282.

Jackson J. & Elgar, M.A. 1993. The foraging behaviour of the Willie Wagtail (*Rhipidura leucophrys*): Why does it wag its tail? *Emu* 93: 284–286.

McFarland D.C. 1984. The breeding biology of the Willie Wagtail (*Rhipidura leucophrys*) in a Suburban woodlot. *Corella* 8: 77–82.

White-throated Treecreeper (*Cormobates leucophaeus*)

Higgins, P.J. ; Peters, J.M. & Steele W.K. Eds. *Handbook of Australian, New Zealand & Antarctic Birds.* Vol. 5.

White-winged Chough (*Corcorax melanorhamphos*)

Boland, C.R.J. et al. 1997. Deception by helpers in cooperatively breeding White-winged Chough and its experimental manipulation. *Behav. Ecol. Sociobiol.* 41:251–256.

Boland, C.R.J. 1998. Helpers improve nest defence in cooperatively breeding White-winged Choughs. *Emu* 98:320–324.

Heinsohn, R.G. 1991. Kidnapping and reciprocity in cooperatively breeding White-winged Choughs. *Anim. Behav.* 41:1097–1100.

Heinsohn, R.G. 1991. Inter-group ovicide and nest destruction in cooperatively breeding White-winged Choughs. *Anim. Behav.* 41:1856–1857.

Heinsohn, R.G. 1991. Slow learning of foraging skills and extended parental care in cooperatively breeding White-winged Choughs. *Am. Nat.* 137:864–881.

Heinsohn, R. 1995. Hatching asynchrony and brood reduction in cooperatively breeding White-winged Choughs, (*Corcorax melanorhamphos*). *Emu* 95:252–258.

Heinsohn, R.G. 1995. Raid of the red-eyed chick-nappers. *Natural History.* February 44–50.

Heinsohn, R.G. 1997. White-winged Choughs. *Nat. Aust.* Autumn:26–31.

Heinsohn, R.G. et al. 2000. Coalitions of relatives and reproductive skew in cooperatively breeding White-winged Choughs. *Proc. Soc. of Lond. B:* Vol. 267:243–249

Rowley, I. 1965. White-winged Choughs. *Aust. Natural History* 15:81–85.

Tuttle, M. and Pruett-Jones, S. 1996. White-winged Choughs, *Corcorax melanorhamphos,* using a stick nest. *Emu* 96:207–208.

Understanding context

Black and white birds
Chisholm, A.H. 1934. *Bird Wonders of Australia.* Angus and Robertson. Sydney.

Woodcock, L., et al. 2004. Mate choice in Black-capped Chickadees: female preference is black and white. American Ornithologists' Union meeting, Université Laval, Québec, QC, Canada, 16–21.

Zahavi, A & Zahavi, A. 1997. *The Handicap Principle.* Oxford Univ. Press. Oxford.

Birds and Willie-Willies
McNaught, R.H. & Garradd, G. 1992. On galahs and vortices. *Emu* 92:248–249.

Garradd, G. 1994. Express elevators to the sky. *Aust. Nat. Hist.* 23: 9–10.

Brightly colored rail chicks
Krebs, E.A. & Putland, D.A. 2004. Chic chicks: the evolution of chick ornamentation in rails. *Behav. Ecol.* 15:946–951.

Lyon, B.E., et al. 1994. Parental choice selects for ornamental plumage in American Coot chicks. *Nature* 371:240–243.

Booming in cassowaries
Mack, A.L. & Jones, J. 2003. Low frequency vocalizations by cassowaries (*Casuarius*). *AOU* conference.

Building mud nests
Rowley, I. 1970. The use of mud in nest-building a review of the incidence and taxonomic importance. *Ostrich Supp.* 8:139–148.

Coping with desert conditions
Ambrose, S.J. 1984. The response of small birds to extreme heat. *Emu* 84:242–243.

Ambrose, S.J. & Wolfgang, J. 1998. Some like it hot: Australian desert birds. *Wingspan* 7:6–9.

Davies, S. 1982. Behavioural adaptations of birds to environments where evaporation is high and water is in short supply. *Comp. Biochemical. Physiol. 71A:* 557–566.

Evans, S.M. et al. 1988. Drinking skills in estrildid finches. *Emu* 89:177–181.

Finlayson, H.H. 1932. Heat in the interior of South Australia and in central Australia: holocaust of bird life. *South Aust. Ornithol.* 158–163.

Fisher, C.D. 1972. Drinking patterns and behavior of Australian desert birds in relation to their ecology and abundance. *Condor* 74:111–136.

167

Maclean, G.L. 1976. A field study of the Australian Dotterel. *Emu* 76:207–215.

Oosterzee, P.V. 1993. *The Centre: the Natural History of Australia's Desert Regions*. Reed, NSW.

Cooperative breeding

Arnold, K. 2000. Strategies of the cooperatively breeding Noisy Miner to reduce nest predation. *Emu* 100:280-285.

Cockburn, A. 1998. Evolution of helping behaviour in cooperatively breeding birds. *Annu. Rev. Ecol. Syst.* 29:141–177.

Heinsohn, R. G. & Legge, S. 1999. The cost of helping. *Trends in Ecology and Evolution*, 14, 53–57.

Koenig, W. D., et al. 1992. The evolution of delayed dispersal in cooperative breeders. *Quarterly Review of Biology*, 67, 111–150.

Dunn, P. O., et al. 1995. Fairy-wren helpers often care for young to which they are unrelated. *Proceedings of the Royal Society of London, Series B* 259:339–343.

Whittingham, L. A., et al. 1997. Relatedness, polyandry and extra-group paternity in the cooperatively breeding White-browed Scrubwren (*Sericornis frontalis*). *Behavioral Ecology and Sociobiology*, 40, 261–270.

Putland, D. A. 2001. Has sexual selection been overlooked in the study of avian helping behaviour? *Anim. Behav.* 62, 811–814.

Zahavi, A. 1990. Arabian Babblers: the quest for status in a cooperative breeder. In: *Cooperative Breeding in Birds.* Stacey,P. B.& Koenig. D.Eds pp. 105–130. Cambridge: Cambridge University Press.

Boland, C. R. J., Heinsohn, R. & Cockburn, A. 1997. Deception by helpers in cooperatively breeding White-winged Choughs and its experimental manipulation. *Behavioral Ecology and Sociobiology*, 41, 251–256.

Clarke, M. F. 1984. Cooperative breeding by the Australian Bell Miner (*Manorina melanophrys*) Latham: a test of kin selection theory. *Behavioral Ecology and Sociobiology*, 14, 137–146.

Corroberees

Rooke, I.J. 1979. The social behaviour of the honeyeater (*Phylidonyris novaehollandiae*) Ph. D. Thesis, University of Western Australia, Perth: RAOU microfiche No. 3, Melbourne.

Do birds play?

Bekoff, M. & Byers, J.A. 1998. *Play Behaviour in Animals*. Cambridge Univ. Press. Cambridge, U.K.

Chisholm, A.H. 1965. *Bird Wonders of Australia*. Angus and Robertson. Sydney

Ficken, M. 1977. Avian play. *Auk* 94:573–582.

Klapste, J. 1980. Rainbow Bee-eater: playful behaviour, and other observations. *Aust. Bird. Watcher* 8: 253–253.

Pellis, S.M. 1981. A description of social play by the Australian Magpie (*Gymnorhina tibicens*) based on Eshkol—Wachman notation. *Bird Behav.* 3:61–79.

Sharland, M. 1971. Chough plays with pine cone. *Aust.Bird Watcher* 4:27–27.

Watson, D.M.1992. Object play in Laughing Kookaburra, (*Dacelo novaeguineae*). *Emu* 92:106–108.

Drinking habits of desert dwelling birds

Fisher, C.D. 1972. Drinking patterns and behavior of Australian desert birds in relation to their ecology and abundance. *Condor* 74:111–136.

168

Duetting

Arrowood, P.C. 1988. Duetting, pair bonding and agonistic display in parakeet pairs. *Behaviour* 106: 129–156.

Farabaugh, S.M. 1982. The ecological and social significance of duetting. In: *Acoustic Communication in Birds*. Eds. Kroodsma, D.E. & Miller, E.H. Academic Press, New York.

Langmore, N.E. 1998. Function of duets and solo songs of female birds. *Trends in Ecology and Evolution* 13: 136–140.

Maller, C.J. & Jones, D.N. 2001. Vocal behaviour of the common koel, (*Eudynamys scolopacea*), and implications for mating systems. *Emu* 101: 121–128.

Striedter, G.F. et al. 1999. Behavioural and neural mechanisms of vocal imitation in adult Budgerigars. *26th International Ethological Conference was held in Bangalore India..*

Tingay, S. 1974. Antiphonal song of the Magpie-lark. *Emu* 74: 11–17.

Wickler, W. 1980. Vocal dueting and the pairbond. I. Coyness and partner commitment. A hypothesis. *Z. Tierpsychol.* 52:201–209.

Egg tossing and mate sharing in moorhens
See Dusky Moorhen

Escape tactics
Lima, S. L. 1993. Ecological and evolutionary perspectives on escape from predatory attacks: a survey of North American birds. *Wilson Bull.* 105:1–215.

Finding freshwater
Vestjens, W.J.M. 1986. Collection of water by the Australian Pelican. *Bird Obs.* 651:32.

Tool use
Grant, P.R. 1986. *Ecology and evolution of Darwin's Finches*. Princeton Univ. Press, Princeton, N.J.

Hawke, D.J. 1994. Seabird association with Hector's Dolphins and trawlers at Lytelton Harbour mouth. *Notornis* 41:206–209.

Higuchi, H. 1986. Bait-fishing by the Green-backed Heron, *Ardeola striata*, in Japan. *Ibis* 128:285–290.

Higuchi, H. 1988. Bait-fishing by the Green-backed Heron in south Florida. *Florida Field Nat.* 16:8–9.

Hobbs, J.N. 1971. Use of tools by the White-winged Chough. *Emu* 71:84–85.

Hunt, G.R. 1996. Manufacture of hook-tools by New Caledonian Crows. *Nature* 379:249–251.

Lefebvre, L. 1995. The opening of milk bottles by birds: Evidence for accelerating learning rates, but against the wave-of-advance model of cultural transmission. *Behav. Proc.* 34:43–54.

Lefebvre, L. et al. 1997. Feeding innovations and forebrain size in birds. *Anim. Behav.* 53:549–560.

Lefebvre, L. et al. 1998. Feeding innovations and forebrain size in Australasian birds. *Behaviour* 135:1077–1097.

Richards, B. 1971. Shrike-tit using twig. *Aust. Bird Watcher* 4:97–98

Roberts, G.J. 1982. Apparent baiting behaviour by a Black Kite. *Emu* 82:53–54.

Stokes, T. 1967. Crimson Rosella catches insects on the wing. *Emu* 66:371.

Townsend, S. 1972. The ingenuity of the butcherbird. *Aust. Bird Watcher* 4:237–238.

Wheeler, R. 1946. Pacific Gulls and mussels. *Emu* 45:3.

Function of flash marks

Brooke, M. De L. 1998. Ecological factors influencing the occurrence of flash marks in wading birds. *Func. Ecol.* 12:339–346.

Stawarczyk, T. 1994. Aggression and its suppression in mixed-species wader flocks. *Ornis Scandinavica* 15:23–37.

Grass-finches

Burley, N.T & Symanski, R. 1998. "A taste for the beautiful": latent aesthetic mate preferences for white crests in two species of Australian grassfinches. *Am. Nat.* 152:792–802.

Immelmann, K. 1959. Experimentelle untersuchungen über die biologische bedeutung artspezifischer merkmale beimzebrafinken (*Taeniopygia castanotis* Gould). *Zool. Jahrb. Abt System*, 86:438– 592.

Immelmann, K. 1962. Vergleichende beobachtungen über das verhalten domestizierter zebrafinken in Europa und ihrer wilden stammform in Austalien. *Z. Tierzücht.* 77:198.

Derek, G. 1982. *Estrildid Finches of the World.* Oxford University Press. Oxford, UK.

Langmore, N.E. & Bennett, A.T.D. 1999. Strategic concealment of sexual identity in an estrildid finch. *Proc. R. Soc. Lond.* B 266:543–550.

Swaddle, J.P & Cuthill, I.C. 1994. Female Zebra Finches prefer males with symmetric chest plumage. *Proc. R. Soc. Lond. B.* 258: 267–271.

Insights in magpie behaviour

Jones, D.N. 2002. *Magpie Alert: Learning to Live with a Wild Neighbour.* University of New South Wales Press, Sydney.

Nealson, T.J. & Jones, D.N. (Submitted) Step-parental care in Australian Magpies: support for the mating effort hypothesis.

Rohwer, S. et. al 1999. Step-parental behavior as mating effort in birds and other animals. *Evolution and Human Behavior* 20: 367–390.

Migration patterns in Australia

Chan, K. 2001. Partial migration in Australian landbirds: a review. *Emu* 101: 281–292.

Dingle, H. 2004. The Australo-Papuan bird migration system: another consequence of Wallace's Line. *Emu* 104: 95–108.

Dingle, H. 2005. Bird migration in Australasia. *Interpretive Birding* 6:1 Published online.

Griffioen, P.A. & Clarke, M.F. 2002. Large-scale bird-movement patterns evident in eastern Australian data. *Emu* 102:97–125.

More facts about Australian cuckoos

Brooker, M.G. & Brooker, L.C. 1989. Cuckoo hosts in Australia. *Australian Zoological Reviews* 2:1–676.

Nest associations

Smith, N.G. 1985. Nesting associations. In: *A Dictionary of Birds.* eds. Campbell, B. and Lack, E. *Poyser, Calton* 389–391.

Non-vocal sounds

Armstrong, E.A. 1965. *The Ethology of Bird Display and Bird Behaviour.* Dover, New York.

Davis, W.J. 2000. Sound production: non-vocal sounds. *Interpretive Birding* 1 (3):1-5.

Gillard, E.T. 1969. *Birds of Paradise and Bowerbirds.* Natural History Press, Garden City, New York.

Godwin, D. 1970. *Pigeons and Doves of the World.* British Museum of Natural History. London.

Manson-Barr, P. & Pye, J.D. 1985. Mechanical sounds. In: *A Dictionary of Birds.* Eds. Campbell, B & Lack, E., pp. 342–344. Poyser, Stafffordshire, England.

Wood, G. 1987. Drumming to a different beat. *Aust. Nat. History* 22:199–201.

Predator avoidance in newly hatched chicks

Göth, A. 2001. Innate predator-recognition in Australian Brush-turkey (*Alectura lathami,* Megapodiidae) hatchlings. *Behaviour* 138:117–136.

Saving energy while flying

Andersson, M. & Wallander, J. 2004. Kin selection and reciprocity in flight formation? *Behav. Ecol.* 15:158–162.

Weimerskirch, H., et al. 2001. Energy saving in flight formation. *Nature* 413:697–698.

Fairy-wrens response to predators

Langmore, N.E. & Mulder, R.A. 1992. A novel context for bird song: predator calls prompt male singing in the kleptogamous Superb Fairy-wren, *Malurus cyaneus.* *Ethology* 90:143–153.

Some like it short—tails, that is

Balmford, A. et al. 2000. Experimental analyses of sexual and natural selection on short tails in a polygynous warbler. *Proc. R. Soc. Lond.* B 267: 1121–1128.

Symbolic use of nesting material

Immelmann, K. 1967. *Australian Finches in Bush and Aviary.* Angus & Robertson. Sydney.

Tails advertise quality

Rathburn, M.K. & Montgomerie, R. 2003. Tail coloration indicates male quality in White-winged Fairy-wrens. *AOU* conference.

Torpor in Tawny Frogmouths (*Podargus strigoides*)

Geiser, F. et al. 2000. Daily torpor in Australian Owlet-nightjars, (*Aegotheles cristatus*). SHOC 2000, Brisbane, Australia.

Körtner, G. et al. 2000. Winter torpor in a large bird. *Nature* 407:318.

Why some females are larger than males

Andersson, M. 1994. *Sexual Selection.* Princeton Univ. Press, Cambridge.

Bildstein, K. 1992. Causes and consequences of reversed sexual size dimorphism in raptors: the head start hypothesis. *J. of Raptor Res.* 26:115–123.

Jehl, J.R. & Murray, B.G. 1986. The evolution of normal and reversed sexual size dimorphism in shorebirds and other birds. *Current Ornithology* 2:65–101.

Moskoff, W. 2001. Why some female birds are larger than males: the meaning of reversed sexual size dimorphism. *Birding* 33:254–260.

Olsen, P. 1995. *Australian Birds of Prey: The Biology and Ecology of Raptors.* Univ. of New South Wales Press, Sydney, and Johns Hopkins, Baltimore. ISBN 0-8018-5357-5)

Safina, C. 1984. Selection for reduced male size in raptorial birds: the possible roles of female choice and mate guarding. *Oikos* 43:159–164.

Temeles, E.J. 1986. Reversed sexual size dimorphism: effect on resource defense and foraging behaviors of nonbreeding Northern Harriers. *Auk* 103:70–78.

Widen, P. 1984. Reversed sexual size dimorphism in birds of prey: revival of an old hypothesis. *Oikos* 43: 259–263.

Wheeler, P. & Greenwood, P. 1983. The evolution of reversed sexual size dimorphism in birds of prey. *Oikos* 40:145–149.

Index

Acknowledgements

A book of this type would be impossible without consulting published research. In appreciation of the work by others, references are listed at the end of this book. Much of the content (updated and revised) was originally published in various volumes of the science magazine *Interpretive Birding*.

Also, numerous people have provided material and/or reviewed parts of the book including: Karen Anthonisen Finch, Dr. Linda Mealey, Jeff Davis, Dr. Hugh Dingle, Dr. Stephen Debus, Gail R. Hill, and Dr. Darryl Jones. Any errors that remain, however, are my responsibility. I am especially indebted to Dr. Linda Mealey for her support and valuable insights into animal behavior.

Although most photographs are of wild birds, a few are of captive individuals. I thank the following institutions for allowing us to photograph birds in their collection: Desert Park in Alice Springs, Brisbane Forest Park headquarters in Brisbane and the Australia Zoo in Beerwah.

Drawings and cartoons:
Mike Bishop, Marilyn Rose (MR), Lynda Strathan (LS), Judith Borrick (JB). Wm. James Davis (WJD), Belinda Brooker, Stephen Debus (SD), Paul Johnsgard (PJ), Aremy McCann; selected captions, Karen A. Finch (KAF)

Photographs:
Gail R. Hill (GRH), Wm. James Davis (WJD), David Stowe (DS), Peter Fuller (PF), Roger Potts (RP), Belinda Cannell, Melanie Rathburn

About the author

Wm. James Davis, Ph.D., is a naturalist with special interest in animal behavior and communication. He has conducted research and workshops in North America, Panama, and Australia. While living in Australia, Dr. Davis created *Interpretive Birding*, a successful science magazine. Currently, he divides his time exploring the natural world and investigating stories for the Terra Explorer Project, a concept he launched in 2009.

He also enjoys sharing his adventures and   knowledge with other naturalists via talks, workshops, and informal gatherings.

CPSIA information can be obtained
at www.ICGtesting.com
Printed in the USA
LVIC06n1555260214
375276LV00001B/6

* 9 7 8 0 9 8 2 2 6 5 4 1 3 *